Contents and revision planner

D0184934

Unit 1 Dynamic Planet

Unit 2 People and the Planet

Unit 3 Making Geographical Decisions

Unit 1 **Section A** Dynamic Planet
Chapter 1 Restless Earth
How and why do Earth's tectonic plates move?

The Earth in cross-section
Revised

The Earth is made up of several different layers. The diameter of the Earth is about 13,000 km and the outer layer – the crust – is between 6 km and 60 km thick. The upper part of the mantle and crust are known as the **lithosphere**.

Figure 1 The Earth in cross-section

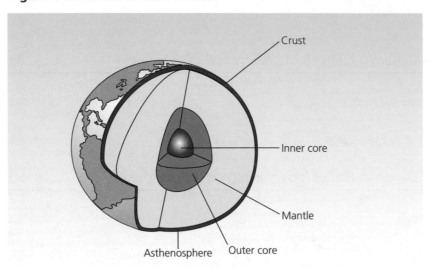

Crust

Inner core

Mantle

Outer core

Asthenosphere

Layer		Physical state	Composition	Temperature (°C)
Lithosphere	**Continental crust**	Solid	Granite	Up to 900
	Oceanic crust	Solid	Basalt	Up to 900
Mantle	**Asthenosphere**	Partially molten	Peridotites	1000–1600
	Mantle	Solid	Silica-based minerals	1600–4000
Core	Outer core	Liquid: very dense	Iron/nickel	4000–5000
	Inner core	Solid: very dense	Iron/nickel	4000–5000

Knowing the basics

There are three basic divisions – the **core**, the **mantle** and the **crust**. The crust is very thin compared with the other two. The core is intensely hot: 4000–5000 °C.

Stretch and challenge

There are many variations within each of the three layers, especially in the mantle. The very top of the mantle behaves like the crust.

Exam practice
Tested

1 Describe two characteristics of the mantle. [2]
2 Describe two characteristics of the crust. [2]

Answers online

The two types of crust

- Continental crust makes up most of the land area of the Earth. It is dominated by rocks that cooled below the surface, such as granite. It is between 25 km and 80 km thick.
- Oceanic crust is much thinner – between 6 km and 8 km thick – and made up of rocks such as basalt.

How the Earth's tectonic plates move

High temperatures in the core caused by gradual radioactive decay create rising limbs of material in the mantle, called **convection currents**. These cool and spread out as they rise before sinking again – just like a lava lamp! Some of this rising and falling material moves in sheets, creating movements in the crust above it, which is pulled apart to form new crust. In other places it rises as columns, creating **hotspots**.

The Earth has a **magnetic field** created by minerals, including iron, that rise and fall in the liquid outer core. This field changes over time.

Figure 2 Convection currents in the mantle

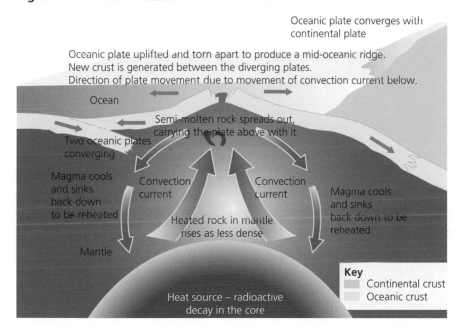

Stretch and challenge

Not all rising material creates movement – in some places the plates move over the hotspots.

exam tip

Make sure that you understand the command word. A question that asks you to *describe* how plates move is not the same as one that asks you to *explain* why they move.

Check your understanding

Place the following in the right order if you undertook a 'journey to the centre of the Earth': core, mantle, crust.

Exam practice

3 What is a 'hotspot'? [2]

4 Explain the causes of tectonic plate motion. [4]

Answers online

Different types of plate boundary

Revised

Knowing the basics

There are three main types of plate boundary:

- **constructive margins**
- **destructive margins**
- **conservative margins**.

Constructive margins

- Constructive margins are formed by rising magma splitting up continental crust and forming new oceans.
- The Eurasian Plate is separating slowly from the North American Plate. The mid-Atlantic ridge is a constructive margin sometimes visible above sea level, as in Iceland.
- This is happening in East Africa today in its continental rift zone.

Destructive margins

- In some places, such as where the Nazca Plate meets the South American Plate, oceanic plates collide with continental plates.
- When this occurs, the denser basaltic oceanic plate sinks beneath the continental plate.
- This process is known as **subduction** and creates a very deep ocean trench near the line of contact between the oceanic and continental plates.
- As an oceanic plate is subducted into the mantle it is subjected to increased pressure and temperature.
- These conditions cause some lightweight materials to melt and rise to the surface to form volcanoes.
- As a result, long chains of volcanoes, known as volcanic arcs, are located above subducted plates, usually above the location where the plate has reached a depth of about 100 km.
- The collision of the plates also lifts and buckles the continental plate, creating **fold mountains**; for example the Andes.

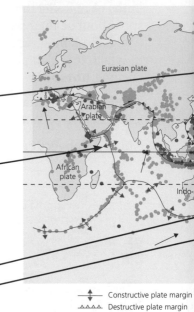

Figure 3 The main plates, margins and volcanoes

↕ Constructive plate margin
▲▲▲▲ Destructive plate margin

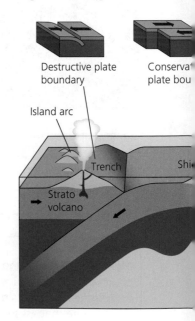

Figure 4 Margins and hotspo[t]

Destructive plate boundary

Conserva[tive] plate bou[ndary]

Island arc

Trench

Strato volcano

Sh[ield]

Exam practice

Tested

5 Using Figure 3, describe and explain the location of volcanoes. [6]

6 Outline one process that takes place at a constructive plate margin. [2]

Answers online

Conservative margins

Where plates slide past each other or move in the same direction but at different speeds then:

- no crust is formed or destroyed, and volcanoes do not form
- great strain builds up along the junction, with sudden lurches along the **fault**
- earthquakes are frequent and often large.

The best known example of this is the system of faults along the west coast of the USA, the best known of which is the San Andreas fault.

Conservative plate margin
Direction of plate movement
• Earthquakes
• Volcanoes

Knowing the basics

Hotspots are areas of rising plumes of magma in the asthenosphere which create volcanoes in the crust as it moves over them, leaving a long trail of island volcanoes.

Knowing the basics

New crust is made at constructive margins and old crust destroyed at destructive margins.

Stretch and challenge

Oceanic crust is created and destroyed. Continental crust is folded, crushed and compressed, but not destroyed.

exam tip

If you are asked to describe a pattern, start with a general point such as 'It is uneven ...'

Check your understanding

1 What are hotspots?
2 What is happening on constructive margins?

Tested

Exam practice

7 Outline the processes that form fold mountains. [4]
8 Explain how subduction zones generate both earthquakes and volcanoes. [6]

Tested

Answers online

Different hazards and their causes

The impact of any **hazard**, including earthquakes and volcanoes, depends on:

- the size of the event
- the vulnerability of the population – poverty and high densities increase vulnerability
- the capacity of the population to cope – how prepared they are.

Different factors influence where volcanoes and earthquakes occur:

- Figure 3 on pages 6–7 shows that volcanoes are not evenly distributed. Because plate margins are essentially lines, so most volcanoes are found in lines.
- **Volcanic eruptions** generate earthquakes, but earthquakes also occur on conservative margins and sometimes happen in regions many thousands of miles from plate margins.
- Not all volcanoes are found on plate margins. Some are found at hotspots where the crust is moving over a column of rising magma (e.g. in Hawaii, see Figure 3).

There are also two main kinds of volcanoes: composite volcanoes and shield volcanoes. This table explains the differences between them. Composite volcanoes are much more dangerous to any human populations nearby.

Type	Form	Magma/lava type	Explosivity and frequency	Example(s)
Composite/ strato	Steep-sided Small area Alternate layers of ash and lava	Viscous/sticky – flows slowly Often 'freezes' in the central vent Granitic or andesitic magma	Infrequent and sometimes unpredictable Pressure builds up over time	Mt Pinatubo (Philippines) Mt Sakurajima (Japan)
Shield	Gentle slopes (like a shield!) Large area Almost all lava	Fluid – flows quickly from many fissures Basaltic magma	Very frequent and generally gentle eruptions	Mauna Loa (Hawaii, USA) Mt Nyiragongo (DRC)

Volcanic eruptions don't generally kill large numbers of people – at least not as a direct consequence of an eruption – because volcanoes are mountains and tend not to be located in areas where lots of people want to live. But many earthquake regions are very attractive areas to live and, despite the risks, some have high population densities. Earthquake events can't be predicted and so are more dangerous than volcanoes.

Large volcanic eruptions and earthquakes can both cause a **tsunami** as a secondary hazard. Tsunamis are a series of very **destructive** ocean waves. Sub-sea earthquakes can displace the seabed and this causes a tsunami. The waves travel across oceans at speeds up to 900 km/h. When they strike land, waves can be 20–30 m high and flood far inland. The Indian Ocean tsunami on 26 December 2004 killed over 250,000 people in fourteen countries.

Exam practice

Tested

9 Describe the differences between composite and shield volcanoes. [4]

10 Describe two characteristics of a tsunami wave. [2]

Answers online

What are the effects and management issues resulting from tectonic hazards?

The impact of earthquakes

A number of factors control the severity of earthquakes:

- the magnitude of the earthquake on the Richter scale
- the depth (shallow earthquakes are more destructive).

In general the impact of earthquakes varies according to:

- the distance from the **epicentre**
- the time of day
- the level of preparedness
- the quality of the emergency services.

For the most part the poorer the country is the greater the impact on people. On the other hand because the population in developed countries insure their property and businesses, the economic cost of the disaster is often higher.

Knowing the basics

Remember that hazards don't necessarily lead to disasters. Much depends on how well prepared people are.

Stretch and challenge

Remember that the size of an event may be too great for even the most prepared countries. The death toll from the Japanese tsunami of 2011 reached over 18,000.

Whatever the impact it is helpful to split it into the following:

1 **Primary impacts** – the immediate effect of an earthquake on property and people. For earthquakes this is the people killed as a result of the shaking and property destruction.

2 **Secondary impacts** – the impact on property and people of an event after it has finished. Lack of shelter and basic supplies, as well as fires, are frequent secondary effects.

Check your understanding

Tested

Learn the key facts for two contrasting earthquake events.

exam tip

Learn some specific facts and figures for your named examples of earthquakes.

Example and details	Earthquake details	Impacts
Port-au-Prince, Haiti January 2010 (developing world)	Magnitude 7.0 Depth 13 km Struck at 5p.m.	**Primary:** 316,000 deaths and 300,000 injured. Total economic losses of £8.5 billion. Poverty and slum housing made people very vulnerable to building collapse and secondary impacts such as cholera. **Secondary:** Over 7000 people killed. An estimated 1 million people were made homeless; damage to roads and ports stopped trade; cholera spread due to lack of clean water and sanitation
Canterbury, New Zealand September 2010 (developed world)	Magnitude 7.1 Depth 10 km Struck at 4.30a.m.	**Primary:** No deaths, about 100 injuries. Total damage to property about £1.8 billion. Deaths were low because most people were asleep and buildings had strong structures. **Secondary:** A major aftershock, magnitude 6.3, occurred in February 2011 in nearby Christchurch, killing 185 people

Exam practice

Tested

11 Outline the differences between the primary and secondary impacts of an earthquake. [2]

12 For a named earthquake, explain its primary and secondary impacts. [6]

Answers online

Living with volcanoes

Volcanoes are **active**, **dormant** or **extinct**. Active volcanoes pose the greatest threat but dormant volcanoes can also be a danger. In 2010, Iceland's Eyjafjallajökull volcano, one of the largest in the country, erupted for the first time in 200 years, disrupting air travel.

> **Knowing the basics**
>
> Volcanoes kill far more people with gas and ash than they do with lava.

In and around Naples in southern Italy over 1 million people live within 9 km of Vesuvius, an active volcano best known for its AD79 eruption, which destroyed Pompeii and other Roman cities.

> **Check your understanding**
>
> Learn the key facts and impacts for a volcanic eruption.
>
> Tested

As with earthquakes the impact of volcanoes can be separated into primary and secondary impacts. In the developed world higher incomes generally mean that the risks of secondary impacts are reduced.

Mount Merapi in Indonesia is a composite volcano that erupted in 2006 and again in 2010. Merapi is a dangerous volcano with a large population living nearby.

Impacts of the 2010 eruption	Social	Economic
Primary impacts	• 360,000 people evacuated from the area; some refused to go and others returned during the eruption • 275 were killed, mostly by scalding hot ash and gas in pyroclastic flows • 570 were injured	• Several villages were destroyed, and damage to crops from ash fall was widespread • About 2000 farm animals were killed • Many flights in the area were cancelled due to the ash cloud
Secondary impacts	• An area 10 km around the volcano was declared a danger zone and 2600 people were not able to return to it • Thousands spent weeks living in cramped emergency centres	• 1300 hectares of farmland were abandoned • Economic losses of $600 million due to severely reduced farming and **tourism** income

Prediction, warning and evacuation

Volcanic eruptions and tsunamis can often be predicted, if the right equipment is in place:

● Gas emissions, earth tremors and 'bulging' of a volcano's flanks can be measured and used to predict eruptions.

● In 1991 about 120,000 people were evacuated from the area around Mt Pinatubo in the Philippines before it erupted.

● Tsunami warning sirens can be used to evacuate people from coasts before tsunami waves strike.

Earthquakes cannot be predicted. However, both preparedness and mitigation can reduce the risk of disaster from earthquakes and volcanoes.

● Preparedness such as emergency plans, well-trained and funded emergency services, warning systems and evacuation routes

● Reducing the impact by mitigation such as hazard resistant buildings, disaster kits and landuse planning.

Exam practice

Tested

13 Define the term 'dormant volcano'. [2]

14 Using an example of a volcanic eruption, describe its impact on people and the economy of the area. [6]

Answers online

An example of mitigation is making buildings safer:

Developed world – make new buildings better	Developing world – make existing buildings better
Foundations very deep but allow movement	Reduce the weight of the roofs
Shock absorbers built into structure	Lightweight, hollow bricks used
Cross bracing to prevent floors collapsing	Strengthen wall corners with wire mesh and cement

Knowing the basics

In the developed world more money is available to make new buildings safer. In the developing world it is important that existing buildings are made safer.

Stretch and challenge

Remember that in the fast growing cities of the developing world new buildings should also include the features often pioneered in the developed world.

Exam practice

Tested

15 Outline two ways that buildings can be designed to cope with earthquakes. [2]

Answers online

exam tip

Remember that if you are asked to outline *one* method of improving buildings to help resist earthquakes then you should offer one point and a development of that point for the two marks available.

Response and relief in Haiti – too little too late
Revised

The event	The response	The analysis
• It was the strongest earthquake in Haiti since 1770 • The 7.0 magnitude earthquake's epicentre was 10 miles west of Port-au-Prince and its 2 million inhabitants • There were many aftershocks ranging in magnitude from 4.2 to 5.9	• Three million people were in need of emergency aid • The Red Cross dispatched a relief team from Geneva, and the UN's World Food Program flew in two planes with emergency food aid • World Vision, an **NGO**, provided food to 1.2 million people, emergency shelter for 41,000 families, delivered 16 million litres of clean water, installed 300 showers and 240 toilets in dozens of camps and were running health, education, child protection and livelihood programmes for tens of thousands of vulnerable children and affected adults • The Inter-American Development Bank immediately approved a $200,000 grant for emergency aid • President Obama promised $100 million in aid to Haiti on 14 January 2009. In comparison, one F-22 fighter bomber costs about $300 million	Haiti is a very poor country and remains so: • Most of the help provided has been from private charities – from NGOs and not from governments • Women are especially badly affected and many are forced into prostitution • The country is still controlled by a tiny elite who may not get aid to the right people. • Even in 2012, Haiti was still trying to recover from the 2010 earthquake

Check your understanding

Explain the difference between preparation and mitigation.

Tested

Exam practice

Tested

16 Describe how the response to the Haiti earthquake attempted to reduce its impacts. [4]

Answers online

Chapter 2 Changing Climate

How and why has climate changed in the past?

Past climate change

The **weather** can change from minute to minute; in the UK it often does!
Climate is defined as the average weather conditions over 30 years.

We know that climate has changed a great deal in the past. This is shown by:

- fossils of animals and plants in regions they are not found in today
- evidence of glaciation in regions that are now free of ice
- evidence from rocks showing us the climate conditions when those rocks were formed
- evidence from ice cores in Greenland and Antarctica showing us how much **carbon dioxide (CO_2)** was in the atmosphere when the ice was formed.

Figure 1 shows a regular pattern of high and low temperatures, like a cycle. This period, known as the **Quaternary**, was a period of rather colder temperatures for most of the time, leading to expanded ice sheets covering much of Europe, Asia and North America.

Figure 1 Temperature change over the past 400,000 years of the Quaternary period

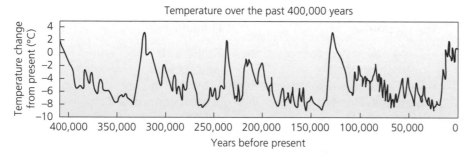

In historical times temperatures have varied by as much as 1.5 °C each side of the average. It only takes very small changes in average temperatures to make a great deal of difference to what you can grow and where you can grow it.

Figure 2 Temperature changes in historical times

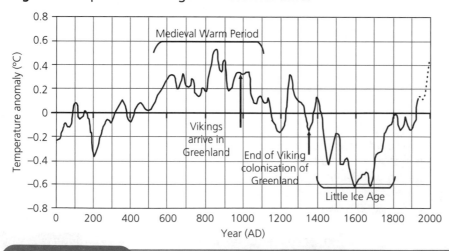

Exam practice

1 Using Figure 1, describe the pattern of temperature changes over the last 400,000 years. [3]

2 Using Figure 2, identify two periods of above average temperatures. [2]

Answers online

The causes of change

There are several theories about past **climate change**. It is possible that these processes operate either together, in which case climate change would be quite severe, or perhaps they 'pull' in opposite directions, in which case the changes would not be as large. The three most important causes are:

1 **Volcanic eruptions**
   - Large eruptions emit vast quantities of dust and gases such as sulphur dioxide into the atmosphere.
   - This blocks out or absorbs incoming solar radiation so the Earth cools.
   - Examples include Mt Pinataubo in 1991, the Laki eruption in 1783 and Mt Toba 70,000 years ago.
   - Large asteroid collisions have a similar cooling effect, as they throw dust and ash into the atmosphere.

2 **Sunspot** activity
   - Sunspots are darker areas on the Sun's surface – they are a sign of greater **solar activity**.
   - They come and go in cycles of about 11 years.
   - However, there are longer periods when very few sunspots were observed, such as 1645–1715.
   - This period coincides with the **Little Ice Age**.

3 Changes in the Earth's orbit and rotation (**Milankovitch mechanism**)
   - The shape of the Earth's orbit changes (becoming more or less circular) over a period of 100,000 years – known as **orbital eccentricity**.
   - The Earth 'wobbles' on its axis over a period of 26,000 years – known as **precession**.
   - The tilt of the axis varies between 21° and 24° over about 40,000 years.
   - Taken together these effects change the amount of solar energy received at the Earth's surface.

Figure 3 The Milankovitch mechanism

Milankovitch cycles drive Ice Age cycles

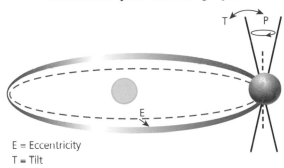

E = Eccentricity
T = Tilt
P = Precession

Check your understanding

Identify three causes of past climate change.

Knowing the basics

All three reasons for climate change can operate together, making the planet hotter or colder, but might also cancel each other out.

Stretch and challenge

There is a lot of evidence that the oceans play a key role in controlling global temperatures and if this gets out of balance then change can accelerate.

Exam practice

3 Explain one way in which volcanic eruptions can cause climate change. [2]

4 Explain the changes in the Earth's orbit that are thought to cause climate change. [6]

Answers online

The Little Ice Age

Revised

One of the best-known periods of climate change in the recent past is the Little Ice Age, probably caused by reduced sunspot activity. It lasted from about 1300 to as late as 1870 (see Figure 2 on page 12) and average temperatures were at least 1 °C below those of today.

Impacts of the Little Ice Age included the following:

● The Baltic Sea froze over in winter, as did most of the rivers in Europe including the Thames.

● Sea ice, which today is far to the north, reached as far south as Iceland.

● Winters were much colder and longer, reducing the growing season by several weeks.

● These conditions led to widespread crop failure and famine.

● Remote areas such as Greenland were abandoned by settlers as survival became impossible.

● The price of grain increased almost everywhere, leading to social unrest and revolt.

● **Glaciers** advanced in the Alps and northern Europe, overrunning towns and farms in the process.

Knowing the basics

Remember that climate change might bring benefits especially if temperatures get warmer in cold regions, such as Iceland.

Check your understanding
Tested

Identify THREE pieces of evidence that the climate might have been colder in the past.

The impact on megafauna

Revised

Megafauna and the Ice Age

The climate change associated with the end of the **Ice Age** some 10,000–15,000 years ago saw temperatures rise by as much as 5 °C in a very short period (by geological standards) of 1000 years.

During this period a number of large animals, so-called **megafauna**, disappeared completely – as many as 130 species in all. Examples include giant beavers, mammoths and sabre-toothed tigers. This is an example of a **mass extinction** event. There have been many of these events in the past, often linked to climate change. There are three sets of explanations:

1 They could not cope with the climate change, with their preferred food – both plant and animal – disappearing; so they too died out.

2 Human beings, our ancestors, hunted them to extinction. We know for sure that we hunted these animals from the evidence of their remains.

3 A combination of 1 and 2 is always possible and maybe most likely.

The 'message' of this is clear. At the moment species are dying out at a faster rate than at any other time in the history of the planet, 1000 times faster than 'normal'.

Knowing the basics

Extinctions take place when species do not have time to adapt.

Stretch and challenge

The most vulnerable species are at the top of the food chain because they depend on the all the other species for their existence. Megafauna are more vulnerable than insects.

exam tip

Some questions will ask you to 'Describe ONE …' but have a two-mark tariff. Make sure that you develop your point rather than introduce a second reason.

Exam practice
Tested

5 Describe the key features of the Little Ice Age. [3]

6 Outline two impacts of the Little Ice Age on people in Europe. [4]

7 State two factors thought to be responsible for the extinction of Ice Age megafauna. [2]

8 State two megafauna species. [2]

 Answers online

What challenges might our future climate present?

The UK has a temperate maritime climate, meaning it is mild and wet. Climate data for Sheffield (below) shows that:

- all months have significant **precipitation**, although May to July are driest
- winter average temperatures, although low, are never below freezing
- summers are warm, rather than hot.

Climate averages for Sheffield 1981–2010:

Month	Jan	Feb	Mar	Apr	May	Jun	Jul	Aug	Sep	Oct	Nov	Dec
Average high °C	6	7	9	12	16	18	21	21	17	13	9	7
Average low °C	2	2	3	4	7	10	12	12	10	7	4	3
Precipitation mm	87	63	68	63	56	67	51	64	64	74	78	92

The UK's climate is unusually mild for its latitude. Moscow and Montreal, on similar latitudes to London, have hotter summers and colder winters. The UK's mild, wet climate is caused by:

- prevailing south-westerly winds bringing moist air from the Atlantic, producing precipitation
- seas surrounding the UK reducing summer temperatures but making winters mild
- the Gulf Stream ocean current that brings warm water across the Atlantic, warming the air surrounding the UK.

Five main **air masses** influence temperatures and precipitation in the UK (Figure 4).

Figure 4 Air masses and the jet stream

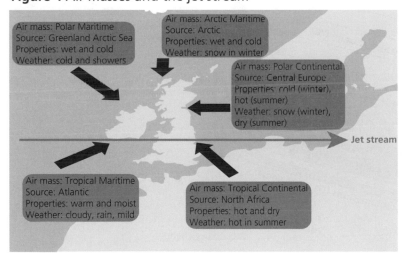

Air masses are large bodies of air with particular properties. They move out of source areas and over the UK, bringing warm, cold, dry or moist air. The jet stream's position influences which air mass the UK experiences at a given time. This is a high altitude wind blowing west to east. If the jet stream moves north, warm air moves over the UK, but if it moves south cold air moves with it.

Future climate change could:

- alter the seasonal pattern of air masses the UK experiences, by changing the jet stream's typical position
- warm the seas around the UK meaning warmer temperatures, but more **evaporation** and rainfall

- make unusual weather extremes more common, such as the summer 2012 floods and summer 2003 heatwave
- disrupt the warm ocean currents that affect the UK, potentially making our climate cooler.

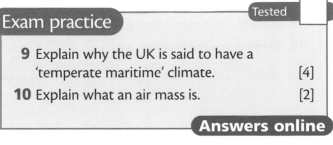

Exam practice

Tested ☐

9 Explain why the UK is said to have a 'temperate maritime' climate. [4]

10 Explain what an air mass is. [2]

Answers online

The greenhouse effect

- **Greenhouse gases** retain heat from the Sun that would otherwise be reflected from the Earth back into space. Without them the planet would be too cold to support life.

- Solar energy passes through the atmosphere without having any real impact on it.

- About half of it is absorbed by the Earth – the rest is reflected back by clouds or the ground, absorbed by clouds or the upper atmosphere or simply scattered back to space.

- But the other half absorbed by the ground is radiated back into the atmosphere and some of this radiation is trapped by greenhouse gases.

Stretch and challenge

The Sun is so hot its energy is all at the short-wavelength end of the spectrum. The cooler Earth re-radiates energy at a longer wave – it is this long-wave radiation that is captured by greenhouse gases.

Greenhouse gases

There are several greenhouse gases, including methane and nitrous oxide, but the one associated most with human activity is carbon dioxide.

Figure 5 Carbon dioxide gas and temperatures over the past 1000 years

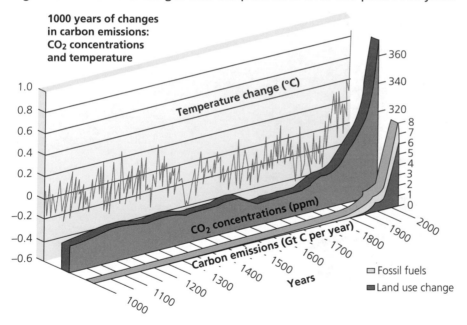
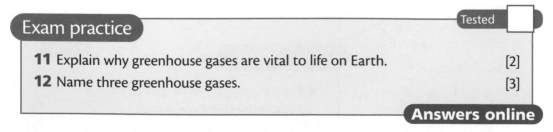

Recent changes in CO_2 are a result of burning fossil fuels such as gas, coal and, above all, oil. CO_2 is also created when making cement and steel. The largest contributing countries are China, USA, Russia, India and Japan.

Exam practice

Tested

11 Explain why greenhouse gases are vital to life on Earth. [2]

12 Name three greenhouse gases. [3]

Answers online

Human activities have increased the amount of CO_2 produced and reduced the ability of the environment to absorb it. Since the **industrial revolution** the levels of all greenhouse gases have risen, contributing to the **enhanced greenhouse effect**.

Year	1850	1950	2012
Carbon dioxide (ppm)	280	323	394
Methane (ppb)*	700	1430	1940
Nitrous oxide (ppm)	270	304	318

*Note that methane is measured in parts per billion (ppb) whereas the others are measured in parts per million (ppm).

The majority of this increase is a result of activities in the developed world where each person produces between 10 and 25 tonnes a year of CO_2 alone. The poorest 25 per cent of people emit less than 2 per cent of the global total. That said, rapid **industrialisation** in India and China is raising their emissions too.

The main reasons for this increase are:

- energy supply (39 per cent of CO_2), which burns coal, gas and oil; most US and Chinese electricity is produced by burning coal
- transport (29 per cent of CO_2), which burns oil; 90 per cent of all journeys are powered by oil
- industry (17 per cent of CO_2) – making things uses energy and produces waste.

Other greenhouse gases have different origins:

- Nitrous oxide is produced by jet engines, fertilisers and sewage farms.
- Methane is most commonly associated with cows producing gas as they graze – about 200 litres of gas per day! With many more people eating meat, cattle numbers have doubled in 50 years.

Human activities also reduce the ability of the environment to absorb greenhouse gases, especially CO_2. The main cause of this is **deforestation**, which has two effects:

1 Burning forest to clear land produces CO_2.
2 Reducing the number of trees lowers the ability of the Earth to absorb CO_2.

Stretch and challenge

Molecule for molecule, methane is about twenty times more powerful as a greenhouse gas compared to CO_2, so rising methane emissions are a concern.

Exam practice

Tested

13 Describe how greenhouse gas levels have changed since 1850. [4]

14 Outline one reason why developed countries produce more greenhouse gases than developing countries. [2]

Answers online

The impact on global temperatures and sea level

Revised

Figure 5 (page 16) showed that global temperatures have risen quite sharply in the past 100 years. The vast majority of experts agree that greenhouse gases are directly linked to global temperature increases. So if we get the relationship between greenhouses gases and **global warming** correct, then models of how things might develop can be drawn up.

As Figure 6 shows, we cannot be certain about the future – these predictions range from an increase of 1 °C if we take action now to 6.4 °C if we do nothing.

Figure 6 Predicted global temperature rise

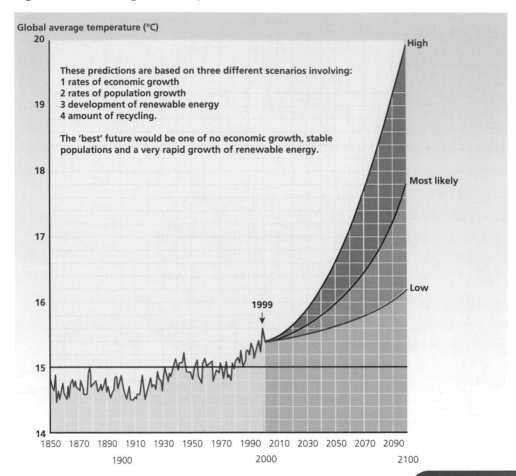

Sea-level change

The most obvious impact of rising temperatures is rising sea level. Sea level has risen by about 200 mm since 1870 (about the width of this page). This has come about for two main reasons:

1 As temperatures rise so water expands – this **thermal expansion** takes place independently of melting ice.

2 Melting **ice caps** and glaciers. Most ice is held in Greenland and Antarctica (99 per cent of all freshwater) – there are 30 million km^3 of ice in Antarctica alone.

There could be a rise in sea level of between 300 mm and 1000 mm depending on how we tackle global warming.

Knowing the basics

The climate has certainly got warmer – the causes are a little more controversial.

Stretch and challenge

Remember that a rise in temperature might lead to other factors accelerating the release of greenhouse gases – for example, by 'unlocking' methane trapped in permanently frozen ground in large areas of Canada and Siberia.

Exam practice

Tested

15 Compare the global temperature predictions shown in Figure 6. [4]

16 Outline one reason why predicting future global temperatures is difficult. [2]

Answers online

Just as past climate change presented both positive and negative challenges, so will future global warming, in both **developed** and **developing countries**:

	Environmental impacts	Economic impacts
The UK (developed)	• New bird and animal species migrate to the UK while others will disappear • Changes to fish species in the sea as temperature rises, e.g. cod moving north out of the North Sea • Increased storminess means more **erosion** on coasts like Holderness	• Cost of protecting low-lying areas such as London and East Anglia from rising sea levels • Costs to the NHS of health problems caused by frequent heatwaves like summer 2003 • More frequent and costly flood events such as summer 2007 and winter 2012 • Domestic tourism could increase but the Scottish skiing industry is likely to disappear • Farmers will need to change crops (from potatoes and wheat to maize and grapes) and irrigation costs could rise
Bangladesh (developing)	• More frequent and/or stronger cyclones in the Bay of Bengal • Rising sea levels erode the country's vital coastal mangrove swamps • Northwest Bangladesh could become more prone to drought	• Flooding becomes more common with increased rainfall and meltwater from Himalayan glaciers, destroying crops and homes • 10% of the land could be lost to rising sea levels, leaving people landless and short of food; 10% of people live less than 1m above sea level • Severe water shortages, if the monsoon rains fail, could lead to widespread famine

exam tip

Be careful with the word 'environment' in questions – it means the *natural* environment.

Exam practice

Tested

17 Using a named example, explain the possible economic and environmental impacts of future climate change in a developing country.

[6]

Answers online

Chapter 3 Battle for the Biosphere
What is the value of the biosphere?

Biomes and their distribution

A **biome** is a global-scale **ecosystem** – that is to say, a community of plants and animals interacting with the non-living physical environment (rock, air and water).

Figure 1 Global biomes

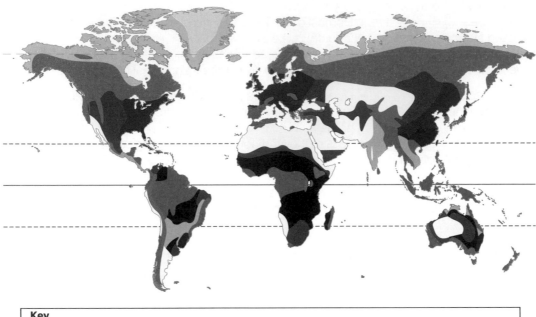

Key

▢	Polar ice	▢	Tundra	▢	Taiga
▢	Mountain zone	▢	Temperate deciduous forest	▢	Temperate evergreen forest
▢	Tropical deciduous forest	▢	Tropical monsoon forest	▢	Tropical rainforest
▢	Shrubland	▢	Temperate grassland	▢	Savannah
▢	Semidesert	▢	Desert		

The main biomes include:

- tropical rainforest found on either side of the equator – a very 'rich' environment where wet and hot conditions encourage all-year-round growth of plants
- desert that is found in the tropics – it is very dry and plants have to adapt to survive with deep roots and thick skins
- temperate deciduous forest found in higher latitudes – trees lose their leaves in the autumn to conserve energy
- taiga (coniferous forest), dominant from 60° northwards for many hundreds of kilometres – it is so cold in winter that trees have **evolved** needle leaves and waxy resin to reduce heat and moisture loss.

Stretch and challenge

Remember that biomes do vary. A tropical rainforest in Malaysia isn't exactly the same as a tropical rainforest in Brazil – the species of plants and animals are very different, although many of the processes are similar.

exam tip

Remember that questions that ask for 'located' or 'named' examples expect you to do just that and preferably not at the 'In Africa ...' scale.

Exam practice

Tested

1 Using Figure 1, describe the distribution of tropical rainforests. [3]
2 Define the terms 'biome' and 'ecosystem'. [4]

Answers online

exam tip

Examination questions will often ask you to 'Describe ONE factor that ...' Make sure that you don't offer two or three factors – you'll only be judged on one of them.

The influence of climate, altitude and soils on biomes

Biomes have developed over very long periods of time. The type of biome depends on differences in:

1 temperature and seasonality
2 rainfall amount and seasonality
3 altitude
4 geology and soils.

Where precipitation is quite high (over 1000 mm) and distributed fairly evenly throughout the year, temperature is the most important factor in biome location. It is not simply a matter of average temperature, but includes other factors such as:

● whether it ever freezes
● length of the growing season (average temperature needs to be above 6 °C for plants to grow).

If there is ample rainfall, we find four characteristic biomes. In order, from the equator, they are: tropical rainforest, temperate deciduous forest, taiga, tundra.

Stretch and challenge

The hotter and wetter the climate, the more productive the ecosystem, with more species of both **fauna** and **flora**.

The other major biomes are controlled not so much by temperature but by the amount and seasonal distribution of rainfall. Rainfall generally reduces as one moves inland away from the oceans.

How much falls and when it falls controls whether the biome will be temperate rainforest, grassland, or desert.

Altitude is considered to be a local factor because it controls both temperature and rainfall. Average temperature falls by about 1°C for every 200 m of altitude.

Soils are also important locally. Thin soils will not support much plant life and slopes and geology will influence this.

Check your understanding

1 Define the term 'tropical rainforest'.
2 Name TWO examples of biomes, other than tropical rainforest.

Exam practice

3 Outline how altitude affects climate. [3]
4 Outline the link between temperature, precipitation and biome type. [4]

Answers online

Biosphere services

The **biosphere** provides many **services** for the planet, acting as a life support system for the other spheres.

Services provided include:

- regulation of climate – temperature and rainfall patterns
- regulation of atmospheric gases – especially levels of CO_2 and oxygen (see Figure 2)
- water regulation through controlling the flow of water
- water purification – water is filtered naturally through plants
- soil formation and development – rotting vegetation is recycled and used continually unless people intervene.

Another set of 'cultural services' is also provided, including:

- spiritual and religious importance, and a sense of place
- recreation, tourism, ecotourism and education opportunities
- inspiration for writers, poets, artists and musicians.

The most important of these is probably the regulation of atmospheric gases.

- Plants absorb CO_2 and produce oxygen.
- Plants use water for growth but also transpire water back into the atmosphere.
- Vegetation regulates temperature.

Figure 2 The carbon cycle

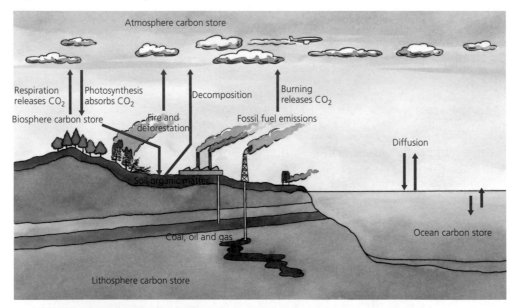

Knowing the basics

The ocean is the largest **carbon store**.

Check your understanding

Outline ONE way in which the biosphere controls the climate.

Exam practice

5 Using examples, explain the value of goods and services provided by a named biome. [6]

6 Explain how the biosphere helps regulate Earth's climate. [4]

Answers online

Stretch and challenge

The ability of the ocean to absorb CO_2 decreases as the ocean warms up.

Goods from the biosphere

The biosphere also produces **goods** for people. Human existence depends upon these goods. Although we modify them and combine them to make new products, all life is dependent on these basic resources.

Resource	Example
Food	All that we eat has its origin in the natural world. 95 per cent of human food comes from 30 plants, and 75 per cent from only eight of those. These crops, such as rice, wheat and maize, are all descended from wild grasses and other plants
Medicines	Quinine is the best-known example of a 'natural' medicine; this treatment for malaria is derived from the cinchona tree. Rosy periwinkle, a plant from Madagascar, gives us two very important cancer-fighting medicines: vinblastine and vincristine. Vinblastine has helped increase the chances of surviving childhood leukaemia from 10 per cent to 95 per cent, while vincristine is used to treat Hodgkin's lymphoma
Raw materials	The most basic raw material is water, the supply of which is seriously threatened in many global regions. Forest biomes are an important source of timber

Check your understanding

Tested

For ONE named biome identify TWO goods and TWO services produced.

Exam practice

Tested

7 Outline one way in which biomes contribute to food supply. [2]

8 Outline one way in which biomes contribute to medicine. [2]

Answers online

How have humans affected the biosphere and how might it be conserved?

Destroying the biosphere – part 1

There is a tension between **economic development** and conserving the biosphere. Economic development in the UK involved major damage to the biosphere as forests were cleared, species wiped out and water and air polluted. Today rapid economic growth in countries such as Brazil is also degrading tropical rainforests.

The Amazon rainforest

Since 1970 about 20 per cent of the Amazon rainforest has been destroyed. The rate of deforestation has slowed recently, but even in 2012 it was 4600 km^2. Figure 3 shows the causes of deforestation. Most forest is cleared for farming, especially beef (cattle) and soy bean farming. The beef and soy beans are exported to the developed world. Logging for tropical hard woods like teak and mahogany is a surprisingly small cause, although one mahogany tree can be worth over US$3000.

The rainforest was 'opened up' from the 1960s by roads such as the Trans-Amazonia highway. This allowed access for loggers and farmers. Large-scale **hydroelectric power (HEP)** dams such as the Belo Monte and Balbina dams have flooded large areas of rainforest. Mines like the world's largest iron ore mine, the Carajas mine, have also destroyed the rainforest.

Small-scale subsistence farmers often clear areas of the Amazon rainforest next to roads. They farm the land for a few years, then move on to clear a new area once the soils become infertile.

Figure 3 Causes of deforestation in the Amazon rainforest

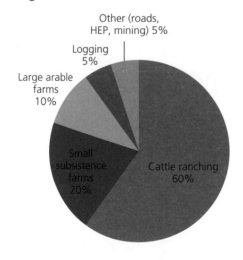

Other (roads, HEP, mining) 5%
Logging 5%
Large arable farms 10%
Small subsistence farms 20%
Cattle ranching 60%

Stretch and challenge

It is too easy to think that destruction of the environment is something that happens in the developing world. It is also worth remembering that deforestation mostly occurs to satisfy demand from rich countries, e.g. for palm oil produced on plantations.

exam tip

The command word 'outline' steers you to give a brief overview and not just to 'name' or 'identify'.

Palm oil

Malaysia and Indonesia are responsible for about 80 per cent of global palm oil production which becomes **biofuel** or is used in the food industry. Deforestation has converted about 6 million hectares of rainforest to palm oil plantation with another 4 million hectares planned by 2015.

Exam practice

9 Using a named example, explain the causes of tropical rainforest deforestation. [6]

10 Outline one reason why people in developing countries find it hard to conserve ecosystems. [2]

Answers online

Destroying the biosphere – part 2

Revised ☐

Human activity alters the biosphere indirectly as well as directly. Global warming has an indirect impact on tropical rainforests:

Figure 4 Indirect impacts on the biosphere

Climate change
- Increased greenhouse gas emissions
- Increases in particles and pollutants

Impacts
- Changes in weather patterns
- Changes in temperature
- Changes in sea level and temperature
- More extreme weather events

Effect on tropical rainforests
- Tropical rainforest dies back as climate becomes arid
- Migration of animals
- Species stress and extinctions

Knowing the basics
Changes in the biosphere are an inevitable result of climate change, especially when the change is rapid.

Stretch and challenge
The damage to the biosphere is obvious – however, in times of economic crisis it scarcely features in government policies.

Exam practice
Tested ☐

11 Describe how global warming could affect the health of tropical rainforests. [2]

12 Outline what is meant by 'conservation' of ecosystems. [2]

Answers online

Biosphere conservation
Revised ☐

There are many examples of attempts at biosphere **conservation**. These range in scale from global to local. Listed below are three examples:

Example and scale	Details
The CITES treaty (global scale)	CITES, the Convention on International Trade in Endangered Species, is a treaty signed in 1973. It protects 34,000 endangered species of plant and animal by banning trade in them. This means it is illegal to trade in elephant tusk ivory, or export rare plants as herbal remedies. It is hoped that CITES will reduce illegal hunting, poaching and plant collection
National parks (national scale)	**National parks** in the UK and USA preserve landscapes. They have strict planning laws that not only aim to preserve valuable landscapes and the biosphere within them, but also allow people to enjoy them. The Lake District was the first UK national park and there are now 15 in total. In the UK the land is privately owned, but in the USA the land in their 59 national parks is owned and managed by the government
UNESCO biosphere reserves (local scale)	Biosphere reserves are used to manage conservation areas at a local scale. They contain three zones: a core protected zone, a buffer zone and an intermediate zone where people live and work. Economic development, education and training take place in the intermediate zone to provide locals with the means to earn a living as well as a way to learn how to protect and enhance the core zone

Exam practice
Tested ☐

13 Describe how one global strategy is attempting to reduce the threat to the biosphere. [4]

14 Outline one local strategy to protect the biosphere. [2]

Answers online

exam tip
Be careful with scales – if you are asked for a local-scale example then make sure that is what you provide.

Being sustainable at a local scale

Across most of the world the biosphere and people exist side-by-side. People want access to forests and other biomes and to use resources from them. To stop the biosphere being permanently damaged by human use, sustainable management is needed. This involves:

- conserving the environment so that it has time to regenerate for future generations
- avoiding exploitation of the area for the profit of people who do not live locally
- provision for local people, especially the poor and disadvantaged
- education of local people so they feel involved in the project as the most important **stakeholders**.

Small-scale areas need careful management in order to be sustainable. An example is the CAZ (Ankeniheny-Zahamena Corridor) forest area in Madagascar. This is a 400,000 hectare protected area in the tropical rainforest. The CAZ project is supported by the Wildlife Conservation Society (an NGO) and the World Bank. Some of the schemes undertaken are shown below:

Some rainforest areas have been given to local villagers so they can manage them; they can use some forest resources to earn money and protect the forest from illegal loggers	The tropical rainforest's 'carbon credits' (ability to absorb carbon dioxide) are being sold to **transnational corporations (TNCs)** and countries in the developed world, so they can meet emissions reduction targets. The money is re-invested in conservation
Alternative incomes such as fish farming in small ponds are being developed, reducing the need to cut down tropical rainforest timber	Education and training are helping farmers learn to farm sustainably in one place, cutting down the need for 'slash and burn' farming
Ecotourism is being developed, with entrance fees re-invested in conservation of the tropical rainforest	Replanting 300 hectares of tropical rainforest. This created jobs for 200 local people

exam tip

Always look at the number of marks awarded. All questions worth less than 6 marks are point marked. One point = one mark.

Exam practice

15 Explain what sustainable management of the biosphere involves. [4]

16 Using a named example, explain how one small-scale forest area is managed sustainably. [6]

Answers online

Chapter 4 Water World

Why is water important to the health of the planet?

Figure 1 Global water supply

Total global saltwater and freshwater estimates

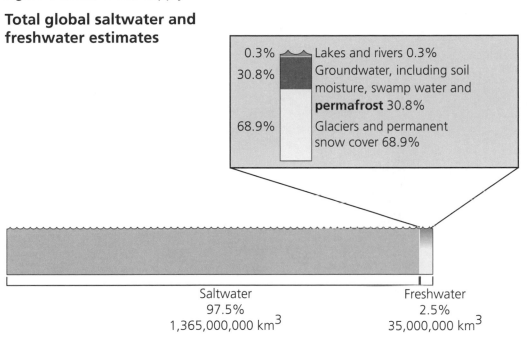

0.3% — Lakes and rivers 0.3%

30.8% — Groundwater, including soil moisture, swamp water and **permafrost** 30.8%

68.9% — Glaciers and permanent snow cover 68.9%

Saltwater
97.5%
1,365,000,000 km^3

Freshwater
2.5%
35,000,000 km^3

Figure 1 shows the water stores on Earth – the so-called **hydrosphere**. Most of this global water is held in the oceans. The other major stores of water are:

- glaciers and snow cover
- groundwater
- lakes and rivers.

Water is part of a closed system. There is a finite amount of water on Earth. The water that you drink has been drunk before, probably many times, and been returned to the system. The water is recycled through a number of processes.

This recycling mechanism is known as the **hydrological cycle**. Water circulates through this via the biosphere (especially plants), **lithosphere** (water is stored in the ground and runs over it and through it) and back to the atmosphere.

Knowing the basics

The vast majority of water is in the oceans. The vast majority of fresh water is held in ice sheets.

Check your understanding

Tested

Identify THREE stores of water.

Exam practice

Tested

1 Using Figure 1, describe the main global stores of water. [3]
2 Explain why the water cycle is a closed system. [2]

Answers online

The workings of the hydrological cycle

Revised

Figure 2 The hydrological cycle

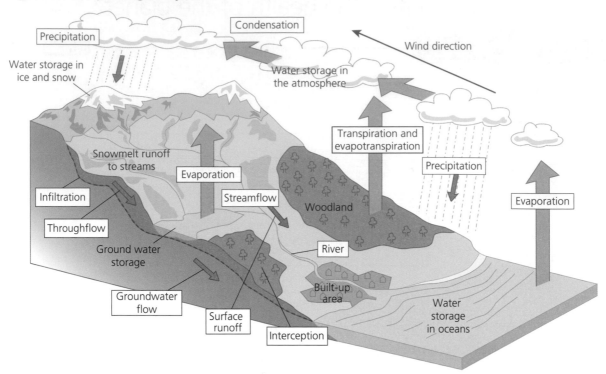

Figure 2 shows the hydrological cycle. This has stores (see Figure 1 for more details) and a set of processes which transfer water from one store to another. Remember that the vast majority of water is held in the oceans.

The key processes are as follows:

- Evaporation – this is the transfer of water from its liquid state to its gaseous state, water vapour. The process depends on temperature – the higher the temperature the more evaporation takes place.

- **Transpiration** and **evapotranspiration** – this is similar to evaporation but involves liquid water lost from plants returning to the atmosphere as a gas. Plants 'breathe out' water but also lose it by evaporation from surfaces such as leaves; hence evapotranspiration.

- **Condensation** – this is where water vapour (a gas) becomes liquid water again. This happens when air containing water vapour is cooled. The result is clouds or, at a low level, fog and mist.

- Precipitation – not all clouds give rain but when the conditions are right, water droplets in clouds (which are microscopic) become water drops or, more often, snowflakes and gravity does the rest – usually the snowflakes melt as they fall through warmer air.

- **Surface runoff** and **groundwater flow** – when precipitation reaches the ground it either soaks into the ground (**infiltration**) and then moves downhill under the surface as groundwater flow or, when the ground is saturated with water or too compact to allow water to infiltrate, it runs off the surface into rivers.

- **Streamflow** – some groundwater flow and surface runoff will reach rivers and streams where it generally continues its route towards the sea or ocean.

> **exam tip**
>
> These processes are important and questions about them are very common. Make sure you know all the processes and in the right order!

Exam practice

Tested

3 State three processes that transfer water around the hydrological cycle. [3]

4 Explain how the hydrological cycle transfers water around the planet. [6]

Answers online

The impact of unreliable and/or insufficient water supply

Human beings are very good at adapting. People can survive and flourish in a wide variety of climates and in environments that seem quite challenging to those of us who are used to expecting water to come out every time we turn on a tap. Knowing that shortages might be coming allows people to change their behaviour to adapt. The most difficult problems arise when:

● shortages are unexpected
● there are long-term changes in both demand and supply.

Some parts of the world are obviously more at risk from water shortages than others. In many arid regions a combination of factors have come together to cause rainfall to become insufficient:

● rising demand because of changing land use – **commercial** farming for example
● rising demand because of population increase – true in both Australia and the Sahel region
● falling supply because of short- or long-term changes in climate – see Figure 3
● falling supply because of neighbouring countries taking more water from rivers.

Figure 3 Changing rainfall in the Sahel region

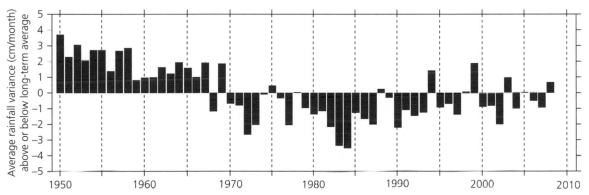

Water in the Sahel

The Sahel region of Africa is a semi-arid region. It is a poor area, where many people are subsistence farmers and **nomadic** herders. Rainfall in the region has been declining, and becoming more unreliable for decades (see Figure 3). This had several impacts on people:

● Severe droughts occurred in 1984–85 (leading to a famine in Ethiopia), 2005 and 2010, reducing food supply as harvests failed.
● Grass and scrub was cleared to allow farming on steep slopes, but drought followed by heavy seasonal rains increased soil erosion, so farming is less productive.
● During dry spells, soil in fields blows away creating dust storms.
● Lower rainfall and eroded land leads to **desertification**.
● As rainfall has decreased and become more unreliable, people have been forced to migrate to cities.

Although water is always in short supply in the Sahel, even the smallest reduction in seasonal rainfall makes the people more vulnerable.

The impact of climate change on vulnerable regions

The impact of climate change is likely to bring more rain to some areas but less to others. In other global regions temperatures might change in such a way as to increase rates of evaporation.

Figure 4 shows some of the changes to **water stress** by 2070. Some areas, such as North Africa, the Middle East and Mexico are predicted to experience extreme water stress (fewer than 1400 litres available per person per day) by 2070 because of rising demand:

● Population increase will mean more thirsty people.

● Increased industry and farming will demand more water.

● Increased wealth means people will use washing machines and have an indoor water supply.

But water supply could also fall, making the situation worse, because:

● global warming reduces rainfall and/or makes it more unreliable

● people use more groundwater than is put back by rainfall, so groundwater declines

● higher temperatures mean more evaporation.

Figure 4 Increasing water stress – the world of water in 2070

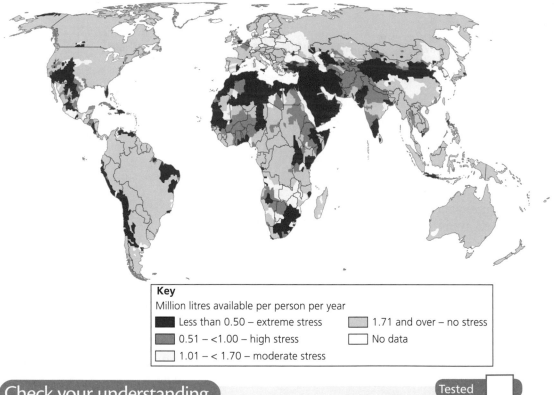

Key

Million litres available per person per year

■ Less than 0.50 – extreme stress
▨ 0.51 – <1.00 – high stress
□ 1.01 – < 1.70 – moderate stress
▨ 1.71 and over – no stress
□ No data

Check your understanding

What do you understand by the term 'vulnerable region'?

Exam practice

7 Using Figure 4, describe the location of areas of extreme water stress in 2070. [4]

8 State two reasons why demand for water will increase in the future. [2]

9 State two ways in which global warming could affect future water supply. [2]

Answers online

How can water resources be managed sustainably?

Figure 5 shows the percentage of the population who cannot access fresh drinking water. This is defined as at least 20 litres of drinking water being available within 2 km of where one lives. It is worth noting that:

- Africa has the worst situation especially in the Sahel region and Central Africa
- there are problems in Asia, e.g. Afghanistan, Cambodia and Mongolia
- even the rapidly industrialising China still has between 18 per cent and 32 per cent without access to fresh water.

Figure 5 Percentage of people who cannot access fresh drinking water

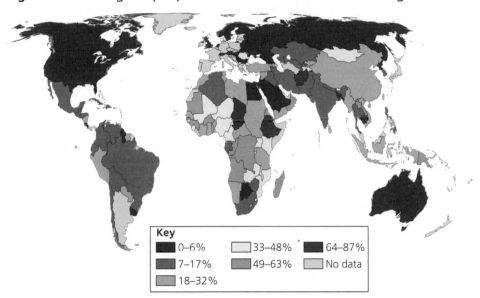

Key

■ 0–6%   ☐ 33–48%   ■ 64–87%
■ 7–17%   ■ 49–63%   ☐ No data
■ 18–32%

Water **pollution** in the developing world is almost always due to untreated sewage in local water supplies. Over 5 million people die each year as a result of water-borne diseases.

Water is also polluted by industrial and agricultural waste, especially in the developed world and **NICs**. Examples are listed in the following table.

> **exam tip**
>
> Questions that ask you to describe distributions on a map expect you to give a 'word picture' of what you are looking at.

Source of pollution	Impacts and example
Toxic waste from mines	Heap-leach mining uses sodium cyanide, which can run off into river water and groundwater, e.g. Fort Belknap (USA). Many US states now ban this method but it is common in NICs
Plant fertilisers	Nitrates run off into rivers and lakes (and the ocean) causing **eutrophication** which kills off animals because they are deprived of oxygen
Chemical waste	Water consumed by people in China contains dangerous levels of arsenic, fluorine and sulphates, often discharged by factories. Of China's 1.3 billion people, 980 million drink water every day that is partly polluted
Radioactive waste	Nuclear waste can remain dangerous for many thousands of years. Storing it deep in the ground is risky because of both the time involved and the chance of leaks

Exam practice — Tested

10 Which areas on Figure 5 have the least available fresh drinking water? [3]

11 State two sources of water pollution. [2]

Answers online

Check your understanding

Identify TWO ways in which water quality might be improved.

Tested

Human interference

Humans can alter the quantity of water supply in a number of ways. This means there is less water available for people, farmers and industry use:

Type of interference	Example and details
Deforestation **Kahama district, north-west Tanzania**	• In Kahama 80% of people are farmers, and deforestation has occurred for timber, land for tobacco plantations, fuel wood/charcoal and **subsistence farming** • Without trees, surface runoff increases and infiltration decreases so groundwater levels in well in Kahama have fallen and rivers flow more seasonally • Major floods have increased because interception by trees has reduced • Cholera and diarrhoea have increased in Kahama because people now struggle to get clean water
Over-**abstraction** of groundwater **India**	• 65% of crop water and 85% of drinking water come from groundwater wells • In some parts of north-west India water tables are falling by 4 m every year, as more water is taken than is returned by rain each year • Farmers constantly pump water using pumps powered by cheap electricity provided by the government • The unsustainable over-abstraction means 60% of groundwater supplies will be in a critical state by 2020

Exam practice

12 State two ways in which deforestation can reduce water supply. [2]

13 Using a named example, explain the problem of over-abstraction of groundwater. [4]

The costs and benefits of large-scale water management projects

Large-scale water management projects, especially dams, are often multi-purpose. This means they have several benefits:

● controlling water flow, by reducing flooding or increasing flow in dry periods

● generating HEP

● storing water in a **reservoir**, which can be diverted for farming, industry or people.

These large expensive schemes do have costs as well as benefits:

exam tip

You could double up here – you have to know a case study of a large water management scheme for sub-topic 7 so why not include brief details for this section too?

Stretch and challenge

The people who gain from large-scale projects may not live nearby, and don't suffer from any of the costs as a result.

Example	Costs	Benefits
Hoover Dam and Lake Mead reservoir, USA Colorado River 1936	• Cost $850 million to build • Almost no water reaches the sea – Mexico's water supply has been reduced • With reduced water flow and no seasonal flooding, the river ecosystem has declined • States and cities argue over who gets the water • Water levels reached a record low in 2010, and global warming could reduce it further	• Lake Mead supplies Las Vegas and cities and farms in California with water • The water supply is clean, low cost and reliable • The lake is a popular place for recreation and fishing • The famous dam is a major tourist attraction • Generates 2 gigawatts of electricity

Example	Costs	Benefits
Three Gorges Dam and reservoir, China Yangtze River 2008	• Cost $26 billion to build • The Three Gorges reservoir is increasingly dirty and polluted by industry on its banks • Over 1 million people had to be relocated as the reservoir flooded • Many ancient monuments were flooded • Some animals, like the Chinese river dolphin, have become extinct due to habitat loss	• Stops devastating flooding on the Yangtze river as in 1954 and 1998 • Generates 22.5 gigawatts of electricity – clean energy (no air pollution) • Despite the dam, the river can be navigated far inland, more easily than before • Water supply to people nearby is more reliable

Exam practice

Tested

14 Suggest two reasons why large-scale water management projects are popular. [2]

15 Using named examples, describe the costs and benefits of a large-scale water management project. [4]

Answers online

Small-scale solutions to water management issues

Revised

Small-scale water management projects are funded by governments as well as by NGOs. They are usually:

● bottom-up projects controlled by the local community
● relatively cheap and easy to set up
● easy to maintain using simple or **intermediate technology**
● addressing local issues, especially water quality.

The table below shows two examples:

Knowing the basics

Most projects have winners and losers; positive things happen but so do negative things.

Example	Explanation and benefits
Rainwater harvesting	• The NGO WaterAid trains villagers in Uganda to build **rainwater harvesting** jars • Large 1500 litre jars are made from local sand, mud bricks, cement and copper pipes • The jar collects and stores runoff from a roof via gutters • Each jar costs about £35 and can supply four families • The water helps families get through the dry season, when rivers run dry and it does not rain
LifeStraw	• The Swiss company Vestergaard Frandsen makes LifeStraw, a plastic tube that uses nano-filtration technology to clean dirty water • LifeStraws come in individual (water is sucked through) and family (water is pumped) sizes • The technology removes 99.9% of bacteria, saving lives

Even these intermediate technology solutions have problems. Rainwater harvesting still relies on rainfall, so if this fails for a long time, water supply runs out. Funding depends on charities and the technical help of an NGO – at least at first. They may be too expensive for some families. LifeStraw is not made in the developing world, only lasts a year, and costs money to buy.

exam tip

Make sure that you get the scale right – if you are asked for a small-scale example then make sure that is what you provide.

Exam practice

Tested

16 State two benefits of using small-scale intermediate technology to improve water supply in the developing world. [2]

Answers online

Section B Small-scale Dynamic Planet
Chapter 5 Coastal Change and Conflict
How are different coastlines produced by physical processes?

Contrasting coasts

Coastal zones are dynamic areas, they change. Coastal zones vary according to:

● the processes that take place – how powerful are the waves?
● the type of rock in the area – how resistant is it to **erosion**?

Coasts are often separated into **hard rock** and **soft rock**.

Stretch and challenge

Remember that the resistance of rock is not just a matter of its hardness – it is also how jointed and fractured it is. A shattered rock, even a hard one, will erode quite quickly.

	Characteristics
Hard rock coasts such as Flamborough Head, Cornwall or Wales	• Cliffs tend to be high and nearly vertical • Cliffs tend to retreat by rockfall, and beaches are made of boulders and pebbles • Landforms such as **caves**, **arches**, **stacks** and **stumps** • **Wave-cut platforms** are exposed as cliffs retreat
Soft rock coasts such as Holderness, Christchurch Bay or North Norfolk	• Cliffs either low, gently sloping or absent • Cliffs retreat by **sliding** and **slumping**, and beaches are made from sand or mud • No stacks or stumps, caves are rare • Wave-cut platforms are unusual

Erosional landforms

Marine processes that erode cliffs and beach material include:

● **hydraulic action** – the force of water striking cliffs and often forcing air into fissures, joints and **faults** so fracturing the rock
● **abrasion** – rock and sand thrown with force against cliffs and dragged by waves across platforms will wear these surfaces away
● **attrition** – the rocks 'thrown' at cliffs and dragged across platforms will themselves be worn down and broken up.

Cliffs are formed by:

● erosion at the base by waves exploiting softer patches of material or fractures in the rock
● the undermining of the cliff-face causing rockfall or slumping
● **weathering** and exposure to rainfall, so material is loosened and gravity removes it.

Exam practice

1 Describe the features of a named hard rock coast. [4]
2 Outline two processes that cause erosion at the coast. [4]

Answers online

Concordant and discordant coasts

Figure 1 The Purbeck coast

This section is **concordant** with the geology running parallel to the coastline. The best known features here are those found at Lulworth Cove – see below.

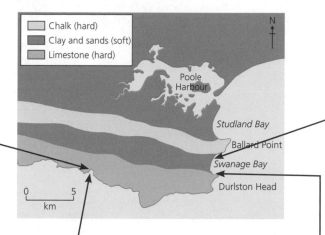

Chalk (hard)
Clay and sands (soft)
Limestone (hard)

Poole Harbour

Studland Bay

Ballard Point

Swanage Bay

Durlston Head

0   5
km

This section is **discordant** with the geology running at right angles to the coastline. The best known features here are those found at Swanage Bay and Studland Bay – see below.

Figure 2 Lulworth Cove

The rocks run parallel to the coast.

At Lulworth coastal erosion has broken through the resistant limestone at A.

Wave erosion has then eroded less resistant sand and clay at B.

Cliffs have been formed at C when a more resistant rock, chalk, has been reached.

Figure 3 Swanage Bay

In this photograph the headland of more resistant chalk can be seen in the distance (D) with its cliffs, and just out of shot, stacks and stumps.

In the foreground (E) is Swanage Bay, eroded from soft clay and sand, as has happened at Lulworth; there are no cliffs but a wide, sandy beach.

Waves can be split into two main groups:

1 **Destructive** or **plunging waves** – these break with a steep descent with little **swash** so **backwash** is strong and erodes material. If plunging waves are close together this can form a rip current that removes a great deal of sand, making the beach even steeper and perhaps creating a **bar** offshore.

2 **Constructive** or **spilling waves** – these spill up the beach (the swash) quite strongly so travel a long distance. Much water soaks into the beach so the returning water (the backwash) is weaker. These waves tend to move sand and other material up the beach towards the land.

Check your understanding

Outline the difference between concordant and discordant coasts.

Tested

Exam practice

Tested

3 State the difference between a concordant and discordant coast. [2]

4 Using Figure 2, explain the shape of the bay at Lulworth Cove. [2]

5 Compare the features of constructive and destructive waves. [4]

Answers online

Waves and landforms

Revised

Figure 4 Contrasting wave types

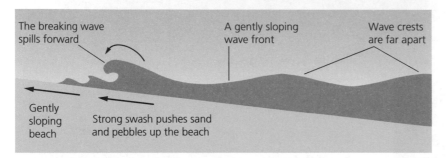

- Steep sloping beach
- The breaking wave collapses
- A steep wave front
- Waves are close to each other
- The backwash pulls sand and pebbles into the sea

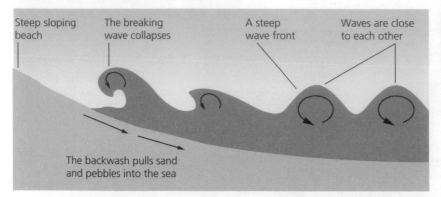

- The breaking wave spills forward
- A gently sloping wave front
- Wave crests are far apart
- Gently sloping beach
- Strong swash pushes sand and pebbles up the beach

Knowing the basics

Check that you understand constructive waves tend to add material to a beach whereas destructive waves remove material.

Check your understanding

Tested

Identify TWO differences between destructive and constructive waves.

Because waves usually approach a beach at an angle, they move material along the beach, as shown in Figure 5.

Figure 5 Longshore drift

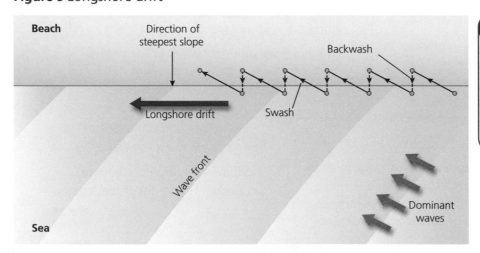

- Beach
- Direction of steepest slope
- Backwash
- Longshore drift
- Swash
- Wave front
- Dominant waves
- Sea

Stretch and challenge

Waves in deep water travel faster than in shallow water. So when waves enter a bay they tend to 'bend' – this process is known as **diffraction**.

1 The dominant wave carries material up the beach as the swash arrives at an angle.
2 However, the backwash returns at 90 degrees to the shore, carrying sediment with it.
3 So material is moved along the beach in a series of swash/backwash movements known as **longshore drift**.

Exam practice

Tested

6 Name the process that moves sand and sediment along a beach. [1]

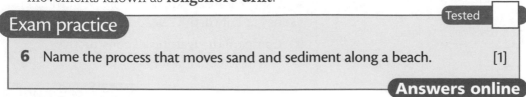

Answers online

How longshore drift forms characteristic landforms

Revised

The best known depositional landform is the beach itself, made up of material eroded from cliffs, brought down to the coast by rivers and moved along the coast by longshore drift.

Figure 6 Formation of spits and salt marshes

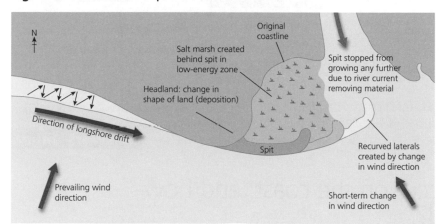

Another landform is the **spit**. As Figure 6 shows, this is formed as follows:

- Material is moved along the coast by longshore drift.
- If there is a change of direction in the coastline this material is deposited across the bay or river mouth.
- In the case of a bay it may eventually be closed off as the sand bar reaches the other side.
- In the case of a river it is likely that the deposition will end when the water becomes too deep and the current too strong.
- Sometimes spits have recurved ends if wind direction varies.
- Behind the spit in slack water **salt marshes** form.

exam tip

Once you understand the process of longshore drift it is quite easy to explain the formation of spits and bars and other depositional features.

Exam practice

Tested

7 Explain the processes that form spits. [6]

8 Which coastal landform is created when a spit closes off a bay? [1]

Answers online

Weathering, mass movement and sea-level rise

Revised

Cliffs are not just eroded by abrasion and hydraulic action. Sub-aerial processes such as weathering and **mass movement** are also important.

exam tip

Most students forget about mass movement and weathering altogether when explaining why cliffs retreat and only talk about waves.

Weathering processes

Weathering is the breakdown of rocks *in situ* (without movement):

- Freeze thaw is a type of **mechanical weathering**; water enters cracks in the rock, freezes, expands and splits the rock apart. Porous and fractured rocks are susceptible to this type of weathering.
- **Chemical weathering** occurs when water reacts with minerals in a rock; it can dissolve the calcite in limestone, or react by hydrolysis with feldspar in granite to produce clay minerals.

Mass movement

Any movement of rock or sediment down a slope, pulled by gravity, is called mass movement.

- During heavy rainfall, cliffs often become saturated with water; water flowing inside the cliff exerts a pressure which weakens the cliff.
- Weak rock layers, such as clay or loose sands in the cliff, can begin to move under this pressure.
- Water often acts as a 'lubricant' inside the cliff which helps gravity pull the cliff down.

Climate change

Climate change could affect coasts in a number of ways. By 2100, sea levels could be 50–100 cm higher than today. The UK may get more powerful winter storms whipping up large waves and bringing more frequent heavy rain. These changes could lead to:

- increased rates of erosion on cliff faces, and more frequent mass movement
- not only erosion of depositional features like spits and sand dunes, but also the deposition of new ones
- flooding of low-lying coastal areas, temporarily during storms, and permanently in some places.

Exam practice   Tested

9 Define the term 'weathering'. [1]
10 Describe how mass movement causes cliffs to retreat. [3]
11 Explain how climate change could affect coastlines in the future. [3]

Answers online

Why does conflict occur on the coast, and how can this be managed?

Rapid coastal retreat – the causes   Revised

Some sections of the UK coast are retreating very rapidly indeed. One of the best known case studies is the Holderness coast in Yorkshire. This is outlined below:

	Holderness
Physical processes	• The coastline is 60 km long with cliffs about 20 m high • The rocks are mostly very weak boulder clay • The dominant waves come from the northeast across the North Sea • Erosion takes place fastest when **spring tides** combine with storms • Rates of erosion are the fastest in Europe at about 2 m per year • The eroded material is carried out to sea so the beaches are narrow and don't act as much of a buffer • The rate of weathering and mass movement is very high because of high rainfall and winter frosts
Problems and threats	• Good-quality farmland is being lost • Many villages have disappeared since Roman times – the land has retreated by over 4 km • Important roads along the coast are at risk, as well as several towns

Knowing the basics

Erosion always causes loss of land. That land varies in value but in the UK that sort of loss is not insured.

Stretch and challenge

Coastal erosion takes place in fits and starts – for most of the time not much is happening although very weak clay is always being weathered and affected by mass movement.

Exam practice   Tested

12 For a named stretch of coastline explain why the coast is rapidly eroding. [6]
13 State two problems for people caused by rapid erosion. [2]

Answers online

How should coastal erosion be managed?

Revised

There are several choices about the management of coastal erosion. The most obvious choice is whether or not the cost of protecting the coast is worthwhile in terms of the savings made for individuals and the country as a whole. The management of sections of coasts in the UK is covered by **Shoreline Management Plans (SMPs)**.

Shoreline Management Plans use one of four options to manage the coast:

- Advance the coastline: move the coastline seaward with land reclamation. This is expensive.
- Hold the line: keep the coastline in the same place, using **hard** and **soft engineering** defences. This is expensive but local people usually want to do this.
- Strategic retreat: allow the coastline to move inland, allowing some erosion but defending when necessary. This creates winners and losers, but is cheaper.
- Do nothing: allow natural erosion and flooding to take place. This is the lowest cost option, but people who live and work at the coast often do not like it.

Check your understanding

Tested

Describe the main purpose of Shoreline Management Plans.

Exam practice

Tested

14 State the meaning of a 'hold the line' coastal management policy. [2]

Answers online

Evaluating defences on the Holderness Coast

Revised

On the Holderness Coast (Figure 7 on page 40) erosion rates vary. This means some places need to be protected from erosion, but others don't. For rapidly eroding areas there are two decisions:

1 Should the place be protected or not? This is often based on the value of the land, buildings and infrastructure.
2 If the place is going to be protected, what should be done? The decision now is often whether to use hard or soft engineering defences.

Hornsea (a small town) has been protected with **groynes**, a **sea wall** and **rip-rap**. Each type of defence has costs and benefits:

Type and purpose	Costs	Benefits
Sea walls £6000–£10000 per metre Protect cliffs from erosion	Can look ugly Restricts access to the beach Large waves can undermine them	Very strong and long-lasting Land and buildings are protected
Wooden groynes £5000 per metre Prevent longshore drift, build up a beach	Need to be maintained, and may only last 10 years Stops longshore drift, so more erosion further along the coast Restricts access along the beach	A large beach is created by deposition, good for recreation The beach dissipates wave energy
Rip-rap £500 per metre Dissipates wave energy	Can look ugly	Absorbs wave energy, reducing the erosive power of waves Looks natural and can be colonised by plants and animals

At Mappleton a combination of hard and soft engineering has been used, including rip-rap, groynes and cliff re-grading. Soft engineering defences look more natural than hard, but are not necessarily cheaper:

Type and purpose	Costs	Benefits
Cliff re-grading Reduce the angle of a cliff, to reduce mass-movement	Expensive and technically difficult Requires other management techniques to prevent erosion at cliff foot	Prevents rapid retreat by slumping or rockfall Makes beaches and shorelines safer
Beach replenishment Pumping sand onto the beach to make it larger	Costs vary, depending on how far the sand needs to be transported Needs to be repeated every few years	A larger beach dissipates wave energy, reducing erosion Looks natural Good for **tourism**

Sea defences at Mappleton and Hornsea have increased erosion further south at Cowden and Aldborough. Groynes trap sand, starving beaches downdrift of their sand so the cliffs are exposed to the full force of waves. This is why **Integrated Coastal Zone Management (ICZM)** has been developed. This approach looks carefully at a long stretch of coast, and decides on a sustainable solution based on:

- which areas to protect and which areas not to protect
- what types of defences to use
- what is best for **ecosystems** and wildlife
- what is best, on balance, for all people on the coast, i.e. not favouring one group more than another.

ICZM means there will be winners and losers. Spurn Head spit is one of the losers. Despite being a unique landform with an important nature reserve and lifeboat station, it is being allowed to erode because the cost of protecting it from erosion is high. Easington Gas Terminal, a nationally important piece of infrastructure, is one of the winners.

Figure 7 The Holderness Coast

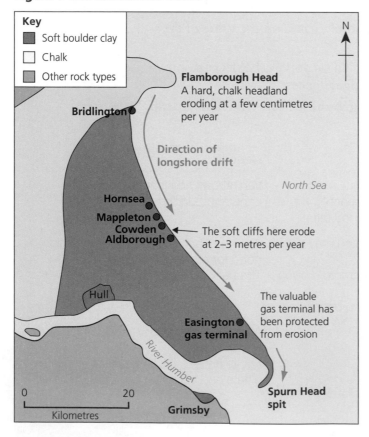

Knowing the basics

It isn't possible to prevent coastal erosion in the long term, but it can be slowed down.

Exam practice

Tested

15 For a named coastline, explain why some locations are protected from erosion but others are not. [8]

16 For a named coastline, explain why hard and soft engineering have been used in its management. [8]

Answers online

Chapter 6 River Processes and Pressures
How do river systems develop?

Contrasts along a river's course

What rivers do:

1 They transport water and sediment from the land to the sea.
2 As a result they are the main mechanism for wearing away (eroding) the land.

No one river has exactly the same characteristics as another but a few generalisations can be made about how rivers change as they move downstream from source to mouth.

Figure 1 The changing course of a river

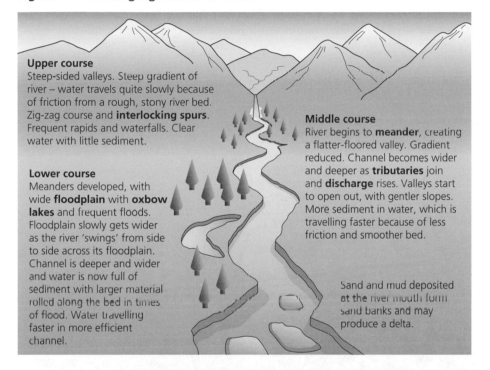

Upper course
Steep-sided valleys. Steep gradient of river – water travels quite slowly because of friction from a rough, stony river bed. Zig-zag course and **interlocking spurs**. Frequent rapids and waterfalls. Clear water with little sediment.

Lower course
Meanders developed, with wide **floodplain** with **oxbow lakes** and frequent floods. Floodplain slowly gets wider as the river 'swings' from side to side across its floodplain. Channel is deeper and wider and water is now full of sediment with larger material rolled along the bed in times of flood. Water travelling faster in more efficient channel.

Middle course
River begins to **meander**, creating a flatter-floored valley. Gradient reduced. Channel becomes wider and deeper as **tributaries** join and **discharge** rises. Valleys start to open out, with gentler slopes. More sediment in water, which is travelling faster because of less friction and smoother bed.

Sand and mud deposited at the river mouth form sand banks and may produce a delta.

Knowing the basics

In most rivers the width, depth and velocity of water flow all increase as you move downstream from source to mouth.

Exam practice

1 Identify two features of a river in its upper course. [2]
2 In which part of a river does a river typically 'meander'? [1]

Answers online

Changes in channel shape and characteristics

The **characteristics** of a river and its channel change along its long profile (Figure 2) from source to mouth. All rivers are different but most fit the general shape of the long profile.

Gradient	Gradient increases downstream, from a steep gradient near the source in hills and mountains to an almost flat gradient at a river mouth
Velocity and discharge	Speed of water flow (velocity) and the volume of water (discharge) increase downstream. Discharge rises because smaller tributaries join the main river flow
Channel characteristics	Channels become wider and deeper downstream. The channel bed gets smoother and the channel is more efficient with less friction between the water and channel sides (which is why velocity increases)
Sediment	Sediment is eroded by attrition so gets smaller downstream, from boulders and cobbles near the source, to pebbles and sand, and eventually silt and clay near the mouth. Total **sediment load** carried by the river increases downstream

Figure 2 River long profiles – ideal and real (River Horner)

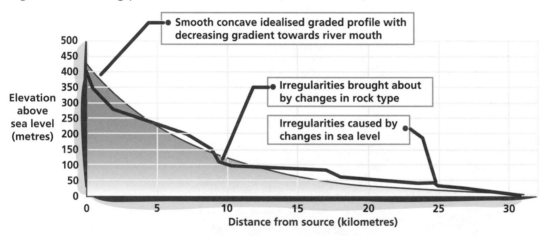

Stretch and challenge

Models of rivers describe how they would develop in a 'perfect' world. In the real world rivers rarely have time to adjust to sea levels rising and falling.

exam tip

Most students imagine that valleys are formed by river erosion and nothing else. They ignore slope processes and weathering.

Exam practice

3 Using Figure 2, describe two differences between the ideal and real long profiles. [4]

4 Describe how river channels and discharge change along the long profile. [4]

Answers online

River landforms and processes

Processes

Rivers erode by:

- hydraulic action – the force of water striking the river bed and banks and often forcing air into cracks and crevices, so fracturing the rock
- abrasion – rocks and particles dragged by water across the bed and thrown against the banks will wear these surfaces away
- attrition – the rocks and particles themselves will be worn down and broken up
- **corrosion** – water will dissolve rocks such as limestone.

Rivers transport their **load** by:

- **traction** – rocks and other particles are dragged along the river bed
- **suspension** – small particles are kept in the water itself until it stops moving
- **solution** – material is dissolved in water, e.g. salts or bicarbonates.

Rivers deposit their load when they slow down as they enter a lake or the sea, or in sections of the river where friction slows down the water.

Typical lower-course features

Figure 3 Floodplains and levées

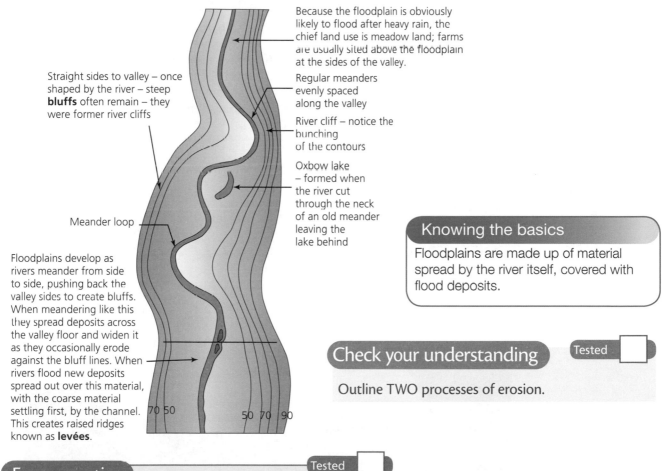

Straight sides to valley – once shaped by the river – steep **bluffs** often remain – they were former river cliffs

Because the floodplain is obviously likely to flood after heavy rain, the chief land use is meadow land; farms are usually sited above the floodplain at the sides of the valley.

Regular meanders evenly spaced along the valley

River cliff – notice the bunching of the contours

Oxbow lake – formed when the river cut through the neck of an old meander leaving the lake behind

Meander loop

Floodplains develop as rivers meander from side to side, pushing back the valley sides to create bluffs. When meandering like this they spread deposits across the valley floor and widen it as they occasionally erode against the bluff lines. When rivers flood new deposits spread out over this material, with the coarse material settling first, by the channel. This creates raised ridges known as **levées**.

70 50   50 70 90

Knowing the basics

Floodplains are made up of material spread by the river itself, covered with flood deposits.

Check your understanding

Tested

Outline TWO processes of erosion.

Exam practice

Tested

5 Describe two erosion processes that occur in rivers. [2]
6 Explain what is meant by the term 'floodplain'. [2]

Answers online

Stretch and challenge

Floodplains are not really flat at all. There are lots of abandoned channels and old levées that make them far from flat.

Typical middle-course features

Figure 4 Meanders and oxbow lakes on the Blackfoot River

Meanders are formed when the faster-flowing water on the outside of the bend (A) erodes whilst on the inside of the bend (B), in the slower-flowing water, deposition takes place. In this way meanders become more obvious – inside bend deposition, outside bend erosion.

Oxbow lakes form when the neck of a meander (C) becomes so narrow that in times of flood the river simply follows gravity and cuts through it, leaving an old meander bend cut off. Deposition soon blocks up the old bend, creating a lake that slowly fills up with material over many centuries.

Figure 5 Meanders in close up

Water flows at different speeds in a channel – the water closest to a bank or a bed is slowed down by friction. Bends occur naturally so as water flows into a bend it keeps on going and, like a car going around a corner, the weight is thrown to the outside of the bend making erosion more likely there – the eroded material is then deposited in the slower-flowing areas further down the channel.

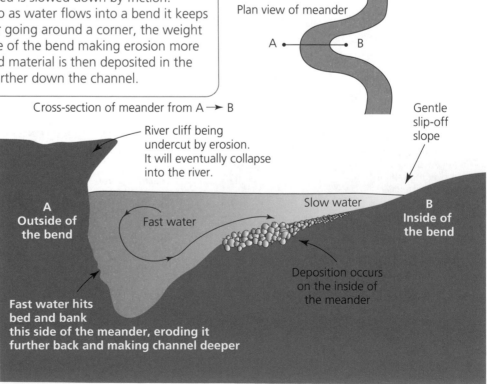

Plan view of meander

A • — • B

Gentle slip-off slope

Cross-section of meander from A → B

River cliff being undercut by erosion. It will eventually collapse into the river.

A
Outside of the bend

Fast water

Slow water

B
Inside of the bend

Deposition occurs on the inside of the meander

Fast water hits bed and bank this side of the meander, eroding it further back and making channel deeper

Typical upper-course landforms

Figure 6 The formation of a waterfall

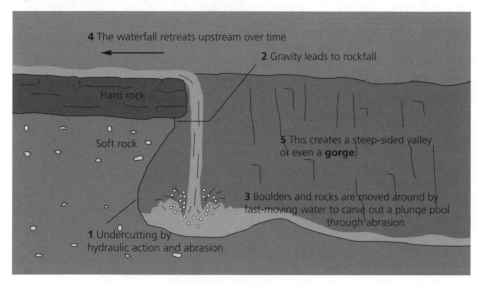

4 The waterfall retreats upstream over time

2 Gravity leads to rockfall

Hard rock

Soft rock

5 This creates a steep-sided valley or even a **gorge**

3 Boulders and rocks are moved around by fast-moving water to carve out a plunge pool through abrasion

1 Undercutting by hydraulic action and abrasion

Knowing the basics

Waterfalls retreat because, over time, erosion and weathering lead the waterfall face to move upstream as it is worn away.

Figure 7 Interlocking spurs and steep valley sides

Steep valley sides (A) are weathered by physical processes such as frost weathering which occurs when water enters cracks and crevices, then freezes and expands, shattering the rock. Mass movement such as rockfall and sliding takes this material down to the stream where it forms part of the load.

Streams cut down vertically. The bed is strewn with rocks and debris which are moved only very occasionally after storms (B). This material wears down the bed and the wandering course of the river creates interlocking spurs as the river cuts down. The river is not very efficient.

Exam practice

Tested

7 Explain how meanders are formed. [6]

8 State the landform which is created when the neck of a meander is eroded. [1]

9 Explain how waterfalls are formed. [6]

10 State two processes which modify the slopes of a river valley. [2]

Answers online

The influence of geology on slope processes

The Grand Canyon is 450 km long and up to 30 km wide, and attains a depth of over a mile – about 1800 m. It is formed in an arid area so weathering of the valley sides is slow, especially on the more resistant rocks. The Colorado River has eroded a deep gorge by abrasion and hydraulic action. Before the river was dammed the discharge would reach over 2500 **cumecs** in spring floods. This is enough to move very large boulders and cut into the channel bed further.

Valley shape

Gorges have near vertical sides because:

- there is little weathering or mass movement
- the river cuts downwards rapidly; perhaps the land is being lifted up too.

In different conditions the valley might be wide because:

- weathering and mass movement operate fast, as in a tropical climate
- rivers erode slowly.

Figure 8 The Grand Canyon

Exam practice

Tested

11 State two processes that transport sediment load in rivers. [2]

Answers online

Stretch and challenge

Rivers do most of their work in rapid bursts of activity when the water levels are high and the flow is rapid after storms or periods of heavy rain.

Why do rivers flood and how can flooding be managed?

Why do rivers flood? Revised

Rivers flood when the amount of water in the channel exceeds channel capacity (**bankfull**) and they overflow. This happens naturally to all rivers from time to time, spreading water and sediment across a floodplain.

Figure 9 A storm hydrograph

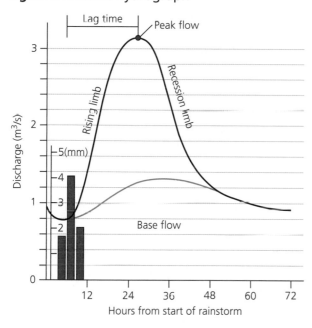

The **storm hydrograph** – a quick guide:

Time is on the *x*-axis starting when the rain event begins.

On the *y*-axis is the amount of water flowing through the channel – its discharge measured in cumecs (m^3 per second).

The blue bar shows the amount of rainfall at different times measured by the inner scale (in mm).

In this example the river water begins to rise a few hours after the rainstorm begins – this is shown as the 'rising limb'.

Eventually, about 26 hours after the start of the storm, the river reaches **peak flow**.

Once this water is passed downstream the discharge begins to fall – the 'recession (or falling) limb'.

The shape of the storm hydrograph depends on many different physical factors. Some hydrographs have long **lag times** before peak discharge, whereas others have short lags and peak rapidly:

Short lag time, high peak Surface runoff and overland flow	Long lag time, low peak Infiltration and throughflow
• Impermeable rocks like clay and granite • Steep slopes • Grassland and thin soils • Short period of very heavy rain • Ground saturated from previous rain	• Permeable rocks, like sandstone • Gentle slopes • Forests with deep soils • Long period of light rain • Previously dry, so ground can absorb the rain

Stretch and challenge

Base flow is the amount of water in the river channel in normal conditions. This water reaches the river by moving underground as **throughflow**.

Check your understanding Tested

Describe the main features of a hydrograph.

Exam practice Tested

12 What is base flow? [2]

13 Explain how physical factors can affect the shape of a storm hydrograph. [6]

Answers online

How people cause flooding

Figure 10 Impact of urbanisation on a hydrograph

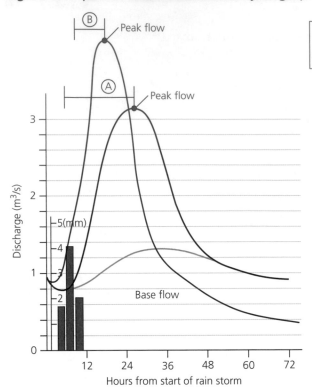

(A) Lag time before urbanisation
(B) Lag time after urbanisation

- **Urbanisation** increases the area of ground where water is going to reach the stream by overland flow because concrete and tarmac are not permeable.
- Drains below the ground will also deliver water quicker than normal soil conditions.
- If water reaches the river quicker then the peak discharge is going to be higher.
- As a result the river will flood more often.

Check your understanding

Urbanisation reduces lag time and increases peak flow. Explain what is meant by this.

There are other ways in which people interfere with what happens in river basins. **Deforestation** removes trees which increases the amount of rain that reaches the ground (leaves intercept rainfall). Trees also use water to grow – removing trees reduces both **evaporation** and **transpiration**, so more water reaches the river, increasing flood risk.

Stretch and challenge

Remember that not all human interference increases flood risk. However, preventing a river from flooding in one place – say by building levées – will simply shift the water downstream so it may make the problem worse elsewhere!

Knowing the basics

Anything that makes ground less permeable (how easily it absorbs water) will lead to more risk of flooding.

Exam practice

14 Outline ways in which human actions can affect the shape of the hydrograph. [4]

Answers online

Traditional flood defences

In 2000, the River Ouse in York peaked at 5.3 m above normal, the highest level since 1625. York railway station on the East Coast mainline was flooded. Three thousand people were evacuated from their homes and 300 properties were flooded (average insurance claims were £25,000 per property). The floods cost York Council £1.3 million, but York lost £10 million as visitors stayed away from the flooded city. In summer 2007, flooding across the UK cost £6 billion and claimed 13 lives. Forty-eight thousand homes and 7,800 businesses were flooded.

Traditional flood defences on the River Ouse, York

Since the floods in 1982, York has been building **traditional (hard) engineering** defences to protect it. These helped reduce the impact of the 2000 floods, and worked again in 2007 and 2012. The **flood defences** are shown on Figure 11.

1 **Clifton Ings.** Open spaces have been left undeveloped to act as a **flood retention basins**, storing over 2 million cubic metres of floodwater, preventing it reaching the city.

2 **Clifton and Bootham.** Schools and houses are protected behind a levée (flood **embankment**). Riverside houses have their own brick and concrete flood walls, and flood gates. In the Marygate area there is a rising flood gate that seals the flood wall at times of high risk.

3 **Riverside apartments** are built with open garages on the ground floor, which are designed to flood. These are easily drained and cleaned, but homes and businesses are safe on higher floors.

4 **The Foss Barrier** is a moveable steel flood gate. During a flood it is lowered into the River Foss, blocking it off. This prevents the River Ouse flooding up the Foss into York. Water from the Foss is pumped over the barrier into the Ouse.

Knowing the basics

Most flood schemes use several different methods to protect people and property.

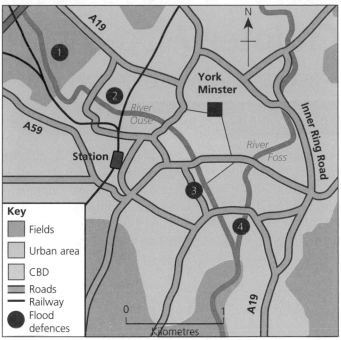

Figure 11 York's flood defences

Key
- Fields
- Urban area
- CBD
- Roads
- Railway
- ● Flood defences

0 ——— 1 Kilometres

Stretch and challenge

Even soft engineering methods have costs – land not built on isn't profitable; forests may not be the most profitable use of land upstream. These indirect costs are rarely included in the calculations.

York's flood defences work well, but they cannot protect against the biggest floods. Some riverside properties flood every few years because defending them would be too expensive. York also has a flood warning system run by the Environment Agency. This alerts people to flood risk.

Soft engineering solutions

Revised

Soft engineering on the River Skerne, Darlington

Hard engineering flood defences are very expensive, can look ugly and impact on river ecology. Dredging a river, i.e. digging a deeper channel to increase channel capacity, destroys the riverbed ecosystem.

A soft engineering approach has been used on the River Skerne in Darlington:

- The once straight channel now has meanders; this means the channel is longer and so can store more flood water.
- 20,000 trees and shrubs have been planted (afforestation), increasing interception and reducing **surface runoff**.
- Wetland marshes have been created. These increase habitats and store water like a sponge during heavy rain.
- **Land-use zoning**: the floodplain is now a park, which is not damaged if the river floods.

These more natural defences and measures are cheaper than hard engineering, but are really only suitable for smaller rivers. They are sustainable as they are cheap to maintain and improve the ecology of the area.

Exam practice

Tested

15 For a named river, explain how flood defences reduce the impacts of flooding. [8]

16 For a named river, explain how soft engineering has reduced flood risk. [6]

Answers online

Section C Large-scale Dynamic Planet
Chapter 7 Oceans on the Edge
How and why are ecosystems threatened with destruction?

A marine ecosystem in detail ⎯⎯⎯⎯⎯⎯⎯⎯⎯ Revised ☐

For this section you will have studied either the **marine ecosystem** of **coral reefs** or mangrove swamps, then looked at its characteristics and its problems. On this page and the next you will find the topic of coral reefs presented in the way that you will need to know for the examination.

Figure 1 Coral reefs classified by threat from local activities

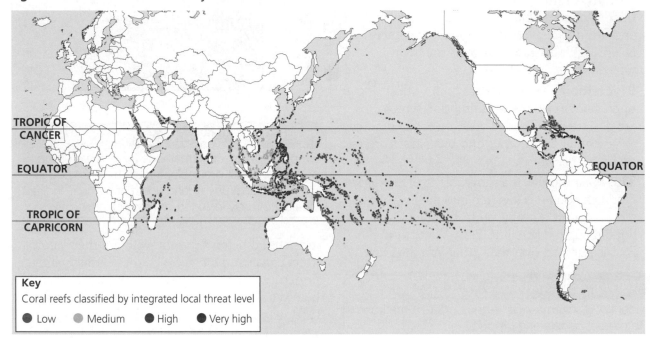

Coral reefs are colonies of living coral organisms. The coral animals, called polyps, secrete a hard calcium carbonate exoskeleton in which the polyps live. Coral reefs also host algae, called zooxanthellae which photosynthesise and provide some of the polyp's food. Coral is found close to the shore in shallow tropical oceans where the sea temperature is between 24 °C and 26 °C (Figure 1).

Coral reefs are not found near major river mouths such as the Amazon and Ganges because the muddy water blocks sunlight so zooxanthellae cannot photosynthesise. This is also why coral cannot grow in water over 25 m deep.

In the last 50 years 20 per cent of coral reefs have been destroyed. Another 25 per cent are under threat of destruction. The major areas where coral has been lost are:

● the Caribbean and Central America
● the islands of Indonesia and the Philippines
● the coastlines in India, Thailand, Vietnam and China
● the Red Sea and Persian Gulf.

Coral reefs are under threat for a number of reasons. This has changed their distribution:

Threats to coral reefs
Global warming Warmer oceans lead to the **bleaching** of coral when the algae that live within it are unable to survive; this upsets the feeding cycle of coral. Local warming can take place following events such as **El Niño**
Fishing Some fishing methods are just plain destructive, such as using explosives (blast fishing). Fishing for aquarium species is also likely to upset the **food web** unless very carefully controlled. Many tropical areas are extremely poor and fish form an important part of the diet – with population increase this is likely to be a major pressure
Coastal development Coastal development almost always leads to an increase in sediment in rivers – disturbed ground, **erosion** of loose soil and the washing off of pollutants all have negative impacts on reefs that thrive in clear water. In some areas reefs provide **lime** for farming – such materials are in very short supply on many tropical islands
Tourism For many developing countries **tourism** is a major income source. Tourists are attracted by reefs but not all of their activities are positive. Not only do activities such as water skiing, scuba diving and surfing cause disturbance, but tourist resorts involve coastal development and increase the pollutants in **lagoons** – what exactly happens to the sewage?

exam tip

Try to avoid extreme statements in your answers. There are many threats to marine ecosystems but some are worse than others.

Exam practice

Tested ☐

1 Using Figure 1, describe the global distribution of coral reefs and reefs at very high threat level. [4]
2 What type of organism is coral? [2]
3 Identify two threats to coral reef ecosystems. [2]
4 For a named marine ecosystem, explain why it is threatened by human activity. [4]

Answers online

How marine food webs are disturbed

Figure 2 A marine food web

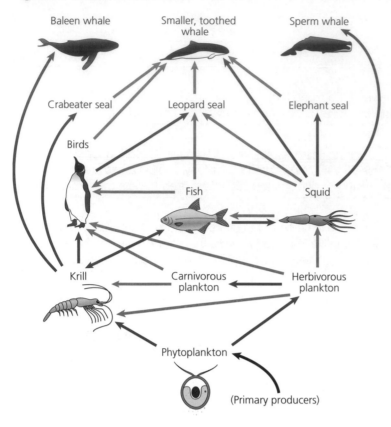

Energy is passed through an ecosystem through a **food web**. This links together the plants (primary producers) to all of the other animals which feed on the plants, and then the animals that feed on those animals. Plants convert energy from the sun into carbohydrate.

Nutrients, such as nitrogen and phosphorus, also pass through marine ecosystems. They move from the physical environment (water, rock and air) into ecosystems and then out again as waste.

Disrupting food webs by overfishing

Food webs are complex so small changes in one element can have large impacts on another.

Large carnivores are generally rare and quickly affected by changes in other parts of the web.

Examples:

1 Whales are still hunted and until recently most whale populations were falling fast.

2 Overfishing can affect both the population of fish themselves and the animals dependent upon them.

3 **Krill** are the cornerstone of this system. Krill are attracting the attention of **commercial** fishing for their protein value.

4 Krill live off **phytoplankton**. Phytoplankton and other aquatic plants and algae consume CO_2 to produce half the world's oxygen output – equalling that of trees and plants on land.

5 Worryingly, the global population of phytoplankton has fallen about 40 per cent since 1950. Global warming is usually identified as the main cause.

Exam practice

5 Define the term 'food web'. [2]

6 Outline two ways food webs can be disrupted by human activities. [2]

Answers online

Eutrophication

The process of **eutrophication** affects marine ecosystems.

Nutrient cycling is critical to all ecosystems. However, when the flow of nutrients into the environment exceeds the ability of natural systems to absorb them, ecosystems feel the impact. The main cause of this excess is fertilisers used on the land and the results are often dramatic:

● Too many nutrients in marine systems can cause excessive growth of algae.
● This blocks sunlight and as it decays removes oxygen from the water.
● This reduces the diversity of species which cannot cope with reduced oxygen.
● This damages coral reefs and other marine ecosystems.
● In extreme cases it creates oxygen-depleted **hypoxic** or 'dead' zones.

Stretch and challenge

Loss of phytoplankton because of global warming is an example of **positive feedback** – oceans get warmer, less phytoplankton grows, so less CO_2 is consumed, so the atmosphere gets warmer, so the oceans get warmer.

Knowing the basics

Most densely populated regions have marine ecosystems that are either eutrophic or, worse, hypoxic.

Siltation

Siltation happens when silt (basically, mud) settles out of sea water onto marine ecosystems. The source of the silt is erosion of farmland, **deforestation** and urban development that increases the volume of silt flowing into rivers and down to the sea. Silt enters the sea via rivers and blocks out sunlight so marine plants such as sea grass, seaweeds and zooxanthellae cannot photosynthesise. It can also smother the seabed and thus destroy the seafloor habitat.

Exam practice

Tested ☐

7   Explain what is meant by the term 'eutrophication'.   [2]
8   Describe how coral reefs can be affected by siltation.   [3]

Answers online

The contribution of climate change

Climate change, in the form of global warming, is causing a number of changes to the oceans. It is likely to accelerate in the future:

Causes	Effects
Warming Ocean temperatures in the tropics have increased by almost 1°C over the past 100 years and could rise 2–3°C more by 2100	**Coral bleaching** As sea temperatures rise, the zooxanthellae algae in coral polyps are expelled and the coral dies. This is called bleaching because the dead coral turns white **Species migration** Cold water-loving species like cod will have to migrate to new areas, which will disrupt fishing as well as food webs
Sea-level rise Sea levels rose by about 7 cm from 1961–2003 and could rise by another 50–100 cm by 2100. This happens because of **thermal expansion** (as the water warms up) and melting ice **glaciers** and **ice caps**	**Changing light levels** Deeper water means that some plant species will no longer be able to photosynthesise and will die out in some places. Coral reefs close to 25 m deep could die out **Coastal flooding** Some coastal ecosystems such as salt marshes, sand dunes and mangrove swamps could suffer increased erosion and permanent flooding
Acidification The oceans are getting more acidic because increased CO_2 in the atmosphere means more is absorbed by the oceans. Between 1750 and 2000, ocean **pH** has decreased from 8.25 to 8.14	**Coral and shellfish health** More acidic oceans could be a problem for marine animals with calcium carbonate shells and skeletons, such as coral and shellfish. Calcium carbonate dissolves in acid water so shells and skeletons will become weaker and harder to grow

The combined impact of global warming and other human threats to marine ecosystems means species extinction is more likely. Many marine species can migrate but coral species cannot do this so easily, and so are more at risk of extinction.

Stretch and challenge

The warming of the oceans is an excellent example of positive feedback. The oceans are warming up because of more CO_2 in the atmosphere but because the oceans can absorb less CO_2 as they warm up, there is more left in the atmosphere, making the oceans warmer and so on.

exam tip

Make sure that you don't muddle up questions that ask you 'why' something has happened and the 'results' of a process. Lots of students know quite a bit about the causes of climate change and sometimes want to show it off, even when the question is about results and not causes.

Exam practice

9 For a named marine ecosystem, explain how it is threatened by climate change. [8]

10 Outline two reasons for current sea level rise. [2]

Answers online

How should ecosystems be managed sustainably?

Investigating the local pressures on a named ecosystem

Marine ecosystems support local populations. In many cases the pressures on these areas are growing. One example is the small Caribbean island of St Lucia, specifically Soufrière Bay which is fringed by coral reefs.

Pressure 1	**Growing population** The island population grew from 130,000 in 1990 to 170,000 in 2010; more people means more people fishing, more waste and more pressure to turn forests into farmland.
Pressure 2	**Tourism** Tourism is the main industry on the island (comprising 45% of **GDP**) and provides income and jobs. The reefs are important – tourists enjoy watersports, go snorkelling, diving and fishing, but yacht anchors can damage the coral, and tourists can leave litter on the reef and remove coral for souvenirs.
Pressure 3	**Runoff and waste** Growing urban areas and tourist resorts mean increased coastal development. Waste such as sewage and farm runoff travels into the sea and eventually damages the coral reef.
Pressure 3	**Fishing** St Lucia is a **developing country** and fish are an important part of the diet. Most fishers do not have the equipment to fish in deep water, so, with a growing population, overfishing on reefs risks damaging the food web.

How different local groups may disagree about management

In Soufrière Bay there are various groups of people who all use the coral reefs differently and have conflicting views:

Group and use of the bay	View	Problems and conflicts
Local fishermen use the reefs as a source of food and income	They want to continue to fish and satisfy the growing demand from local people and tourists	Damaging and unsustainable fishing will reduce fish supply Fishermen and tourists often want to use the same areas of the reef
Tourists use the reef for diving and the sea for watersports	They want a pristine reef, which contains colourful coral and fish species	Different tourists can disagree, e.g. scuba divers and jet skiers High numbers of tourists could damage the ecosystem
Yacht owners and cruise ships anchor in the bay	Soufrière Bay is a sheltered place to anchor and offload tourists to visit the island. They need space to do this	Boat propellers, anchors and waste thrown overboard can all damage the reef Cruise ships and small fishing boats are incompatible. Oil spills are a risk
Local businesses depend on tourism, focused on Soufrière Bay	Shopkeepers, bar owners and hoteliers want tourists, but also want the sea and reef kept natural and pristine to encourage tourists to visit	More tourists could easily mean more damage to the reef and more conflict with fishermen

Exam practice

11 For a named marine ecosystem, explain why there are conflicts over how it should be used. [6]

12 Outline one possible consequence for marine ecosystems of population increase in coastal regions. [2]

Answers online

Local case studies of marine management

Managing marine areas is difficult because of all of the different groups involved. Below are two contrasting examples of management:

Soufrière Bay, St Lucia

- The Soufrière Marine Management Area (SMMA) was set up in 1992.
- The SMMA brought all the different groups together to try and resolve their conflicts.
- The bay was zoned, so that conflicting activities were given their own areas of the bay separate from others.
- Some reef areas are off-limits to everyone, creating marine reserves areas for young fish to grow.
- Locals were trained to police the area.
- Funds from tourist taxes (a day's diving costs £3, anchoring a yacht costs £10) are used to pay for the area's management.

The SMMA has been successful. Damage to the reef has been slowed in most areas, and reversed in the marine reserve areas. However, the pressures from tourism and fishing continue to grow.

North Sea, northern Europe

- The EU Common Fisheries Policy aims to conserve fish stocks.
- Quotas restrict catches and the number of fishing boats is restricted.
- Individual countries like the UK have fishing patrol boats that check the catches of fishing boats.
- In 2013 the EU agreed to reduce by-catch (unwanted fish that are thrown back into the sea, dead), which is estimated to equal 1 million tonnes each year.
- Some 'no-take' zones have been introduced where fishing is banned, such as off the East Yorkshire coast.
- Countries have signed the Convention for the Protection of the Marine Environment of the North-East Atlantic (OSPAR) which aims to reduce **pollution** flowing into the sea via the surrounding rivers.

The North Sea still has problems. Fishing is not yet sustainable, and fewer fish need to be caught to let species like cod, haddock and whiting recover. More, and larger, no-take zones are probably needed. It is very difficult to resolve the tension between the demands of fishers (fishing is their income), consumers (who want cheap fish) and the environment, which needs lower fish catches to be sustainable.

Knowing the basics

Remember that there will always be arguments about using a resource like a reef. Sustainable management involves helping people agree that cooperation is the best route forward.

Exam practice

13 Using a named example, explain how a local-scale marine ecosystem has been managed. [6]

14 State two ways in which the problem of overfishing could be managed. [2]

Answers online

Global solutions to the declining health of the oceans

Revised

Local actions are not enough to protect the oceans. Sixty per cent of the ocean is neither owned, nor managed, by any country. This means some global frameworks are needed to protect the oceans and some of the species in it.

Since the 1980s there has been increasing concern about the growth of 'garbage patches'. The best known is the Great Pacific Garbage Patch. Eighty per cent of its garbage is thought to come from land-based sources and 20 per cent from ships. A typical cruise ship with 3000 passengers produces over 8 tonnes of solid waste every week, much of which ends up in the patch.

The 1983 International Convention for the Prevention of Pollution from Ships (MARPOL) aims to reduce ocean garbage:

● 136 countries have signed MARPOL.

● It bans dumping of waste and oil at sea from ships.

● However, it is hard to police and some countries, for example Thailand, have not signed up.

Individual species can also be protected:

● A ban on hunting whales was brought in by the International Whaling Commission in 1982.

● However, a small number are still caught by a few countries like Japan and Norway.

Marine Protected Areas (MPAs) are like **national parks** for the sea. There are over 5800 globally:

● They protect areas in a similar way to the SMMA on St Lucia.

● However, many MPAs are 'paper parks'; laws are not enforced and money for management is not available.

● Fewer than 1 per cent of coral reefs are protected by effective MPAs.

The oceans are so vast that they are difficult and costly to manage, either locally or globally.

Exam practice

Tested

15 Describe two ways marine ecosystems can be managed sustainably at a global scale. [4]

16 Outline two problems of managing marine ecosystems at a global scale. [4]

Answers online

Chapter 8 Extreme Environments

What are the challenges of extreme climates?

The characteristics of extreme climates

Extreme environments are located in two places:

● **polar regions** such as Alaska and Siberia in the Arctic, as well as the Antarctic
● **hot arid regions** such as the Sahel or central Australia, close to the tropics.

The climate of these two extreme environments is very different:

	Polar	Hot arid
Temperatures	Average monthly temperatures often below zero, and never above 10°C in true polar regions. Winter months can average –20 or –30°C	Long hot summers – average monthly temperatures frequently above 30°C and never below 15°C
Precipitation	Dry – often less than 300 mm per year. Most **precipitation** is snow	Dry – often less than 500 mm per year. Most precipitation is through short but heavy rainstorms
Seasonality	Long cold winters with up to 24 hours of daylight in summer and 24 hours of darkness in winter. Very short summer growing season	No 'winter' as such, but many hot arid areas have a short wet season and a few cooler months. Clear skies so nights can be cold
Variability	Becoming more variable in the Arctic because of climate change, e.g. milder winters. Often very windy with blizzards	Low variability, although periods of extreme drought do occur and dust storms are common

In both cases all species face challenges of extreme temperatures made worse by variability such as windiness or lack of water.

Knowing the basics

Make sure that you know something about temperature and precipitation for both extreme environments.

exam tip

Questions might ask you to describe the climate – try to cover at least some detail in your answer and, if possible, use figures.

Exam practice

1 Compare the characteristics of polar and hot arid extreme climates. [4]
2 Define the term 'extreme environment'. [2]

Answers online

Adapting to the extremes

There are very few environments on Earth in which life cannot survive in some form or other. However, in order to survive, species have **evolved** methods of coping – adaptations that make their survival more likely. In this section we look at a few examples of these adaptations for the **fauna** (animals) and **flora** (plants) that live in extreme climates.

	Hot arid	Polar
Adaptation 1 – Flora	Water shortages have caused many plants to have extensive root systems that both spread out over large areas and go as deep as 40–50 m into the ground to find water	The frozen soil (**permafrost**) has caused many plants to have shallow roots to tap liquid water in the topsoil
Adaptation 2 – Flora	Water is stored in the roots, stems, and/or leaves of plants. The plants that do this are called succulents, often cacti	Plants form rounded 'cushions' 5–10 cm high which protect them from bitter winds
Adaptation 3 – Fauna	Desert animals have adapted to keep cool and use less water. Camels, for example, have large fatty deposits (their humps) that store water. Many desert animals only come out at night to hunt	Many Arctic animals either migrate out of the region in winter or hibernate during the worst of winter
Adaptation 4 – Fauna	Many desert animals are very small with a large surface area to lose heat including features such as big ears	Polar bears, musk ox, reindeer and Arctic foxes all have thick fur to protect from the cold. White fur acts as camouflage in the snow

Both extreme environments are fragile. They cannot cope with much human interference or change (such as global warming). This is because:

● flora and fauna are so closely adapted to the climate, they could not survive if it changed
● there is little vegetation cover to protect against soil erosion
● many animals have specific breeding times and patterns which would be easily disrupted.

Stretch and challenge

Extreme climates pose challenges – there may be ways to adapt but there are fewer plant and animal species in these regions than in less difficult environments.

Knowing the basics

Make sure that you know the difference between flora and fauna.

exam tip

Make sure that you read the question very carefully – it might ask you about plant adaptations or animal adaptations, or both.

Exam practice

3 Describe how species have adapted to the polar environment. [3]
4 Outline two adaptations flora or fauna make to conserve water in hot arid environments. [2]

Answers online

How people adapt

Revised ☐

Extreme environments have low carrying capacities. This means they can only support small human populations because resources are scarce. People can be kept healthy by 'importing' resources from elsewhere, which is why there are towns and cities in hot arid areas and the Arctic, but these communities do not 'live off the land'. To live in an extreme environment means to adapt:

	Hot arid	Polar
Food supplies and farming	Traditionally, **indigenous** people in the African Sahel planted crops in zai pits (a hole with organic matter in it to absorb seasonal rain, covered with soil to prevent **evaporation**). They dig boreholes and wells to trap groundwater	Farming is not possible. Traditionally, seasonal hunting was vital, with food being frozen or salted. The Inuit and Saami adapted to diets based on protein and fat
Building design	Traditional houses have flat roofs (for sleeping outside) and small, shuttered windows in thick walls to keep the heat out. Underground homes and solar panels are modern adaptations	Steep roofs are needed to shed snow, and modern buildings are triple-glazed and face into the sun to maximize solar gain. They are built on stilts to avoid problems of melting the frozen ground (permafrost)
Body shapes and clothing	Many indigenous people like the Masai are tall and slender – an advantage in terms of losing body heat. Clothing is light and loose to allow air circulation	Arctic Inuit are short and stocky, helping to conserve heat. Clothing is multi-layered and uses local skins and furs. Modern 'engineered' fabrics have also been developed
Transport and communications	Camels, which can cope with aridity, were used in the past and travel was often at night. Modern roads and rails have to be designed to cope with heat and resist buckling	In permafrost areas the ground is frozen and only the top few metres ever melt. Pipelines are built above ground on stilts and roads are built on gravel pads to stop the permafrost melting. Most travel is in winter as it is easier on frozen ground
Energy conservation and use	In urban areas, parks and **irrigated** green areas can help reduce the excessive heat, and buildings are carefully designed to reduce the need for air conditioning, e.g. windows face away from the sun	Energy use is very high. It is reduced by very thick insulation, but energy bills are still high. Many polar regions like the Arctic and Siberia actually have large oil and gas fossil fuel reserves, which help reduce energy costs

exam tip

Whichever your chosen climate remember that if you are asked, for example, to 'Outline ONE feature of building design' for two marks, you need to make one basic point and a development of that point for the second mark.

Exam practice

Tested ☐

5 Briefly explain why extreme environments often have low carrying capacities. [2]

6 For either the polar or hot arid extreme environment, explain how people have adapted to live there. [6]

Answers online

Indigenous people in extreme environments have unique cultures. Until recently, many were unconnected to the modern world. In the last 50 years this has changed rapidly, but the close relationship of these people to their natural environment means we can learn from them:

	Polar	Hot arid
Cultural uniqueness – point 1	The ceremonies surrounding whale and seal hunting are part of what gives Arctic communities their cultural identity. Most societies are tribal, with no 'central' leadership, and often live in remote bands	The Tuareg people of North Africa became excellent astronomers, using the stars to navigate the featureless Sahara desert
Cultural uniqueness – point 2	Inuit people were **nomadic**, moving as the herds of caribou moved between winter and summer regions and living in different types of 'houses'. They live in igloos in winter and portable tents made of animal hides and skins in summer	Women have high status in Tuareg society compared to many traditional cultures, for instance, they own the family property and livestock, and are often literate
Values – 1	Inuits treat people, the land, animals and plants with equal respect. They use every part of any animal they catch, wasting nothing. They have strict hunting rules, such as not killing any animal while it is in its mating season	The Tuareg became great traders, moving their camel 'caravans' across the desert, trading salt, food, clothing and animals – trade meant everyone could gain the resources they needed
Values – 2	Traditionally Inuits share the food they have hunted and everyone does his or her part to help those in need – a real community effort	The Tuareg became experts at conserving water, because it was in such short supply – something modern peoples could learn from

Stretch and challenge

Indigenous peoples are illiterate by western standards of reading and writing but they have always had a deeper understanding of the natural world because they live so close to it.

Knowing the basics

Remember that 'culture' is quite a broad term concerning the traditions and lifestyles of particular groups of people.

Exam practice
Tested ☐

7   Outline two traditional activities carried out by people living in polar environments.   [2]

8   Explain why the culture of traditional people in hot arid areas should be valued.   [4]

Answers online

How can extreme environments be managed and protected from the threats they face?

The threats to people and natural systems

The threat to the lifestyles of indigenous peoples and the natural systems in extreme environments has increased very rapidly in the past 50 years. Often the threats are related to increased accessibility of once isolated places:

	Polar	Hot arid
Out-migration	In the Arctic there are jobs in the oil, gas and mining industries, but the work is physically hard and the high pay cannot compensate for the **isolation** and lack of social opportunities	In the Sahel, **subsistence farming** is affected by drought. Basic human needs like food and clean water are often not met so people migrate if they can
Cultural dilution	In Alaska 'western' industry and immigration by oil workers has eroded traditional culture. Nomadism has declined, diets have changed, and old storytelling traditions have been lost. Alcohol has often had a very damaging effect on communities	Masai tribespeople in Kenya and Tanzania often perform dances and sell souvenirs to tourists on safari; this is a 'watered down' version of their traditional culture, one that 'sells'. Contact with tourists increases preferences for meat diets and alcohol
Resource exploitation	Oil, gas and ore drilling and mining have changed much of the Arctic landscape in Alaska and Siberia. The wilderness is criss-crossed by roads and pipelines, disrupting animal migration routes	Much of the land in the Sahel has been over-grazed by cattle and ploughed up for crops. This is often very 'marginal' land that cannot support farming for more than 1–2 years, but population pressure means it has to be farmed
Land degradation	In Siberia, oil spills are common. Up to 5 million tonnes leak every year from rusty pipes and old wells. About 0.5 million tonnes leak into Siberia's rivers, damaging river ecosystems	**Desertification** turns farmland into desert. In the Sahel, it is a serious problem; poor farming practices clear too much vegetation on steep slopes, and heavy seasonal rains cause massive soil erosion

Global warming will make many of these threats worse:

● A warmer Arctic will mean a more accessible Arctic, with more oil, gas and ore exploitation, and more industry.

● In the Sahel, and other hot arid regions, a hotter drier climate will worsen desertification and force more people off the land.

Stretch and challenge

Remember that it is widely thought that changes to the environment caused by global climate change will make these impacts more rapid and more pronounced.

Exam practice

9 Outline two ways in which the culture of polar environments is threatened. [2]

10 For either a polar or a hot arid environment, consider the threats to its natural environment. [8]

Answers online

The threat of climate change

In this section the threat of climate change is assessed. In extreme climates plants and animals are adapted to a narrow range of difficult conditions – quite small changes in the environment lead to large changes in the ecosystems.

For hot arid regions changing rain patterns are the main result of climate change. Figure 1 shows the changing size of Lake Chad from 1963 to 2001, which has reduced supplies of fresh water for drinking and irritation in the Sahel.

Figure 1 The disappearance of Lake Chad

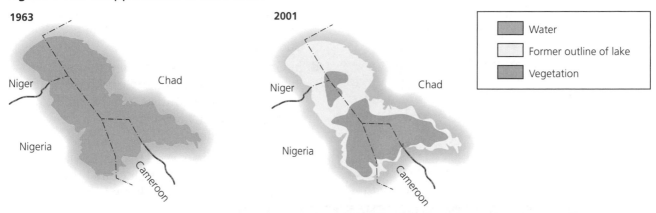

Climate change will have a wide range of impacts on extreme environments:

Hot arid	Impact on natural systems	Impact on traditional economies
Lower/more variable rainfall and higher temperatures	• Increased desertification and erosion of topsoil • More heavy rainstorms, causing flooding, erosion and landslides • Species forced to migrate, and disruption to traditional grazing and breeding areas • Increasing numbers of dust storms	• Increased frequency of drought and famine, and falling human health due to malnutrition • Forced **migration** to find grazing land as traditional areas turn to deserts • **Rural–urban migration** because of falling food and water supply
Polar	**Environmental impact**	**Impact on people**
Warmer temperatures, milder, shorter winters and more storms	• Melting permafrost as temperatures rise • Loss of Arctic sea ice, so ice-dependent seals and polar bears cannot feed and breed • The boreal forest treeline moves north and the tundra grassland area shrinks • Increased coastal erosion rates	• Traditional hunting lifestyle undermined as prey becomes more scarce; diets have to change • Migration to urban areas as traditions disappear and communities split up • Threat to coastal communities due to erosion

Exam practice

11 What is meant by 'desertification'? [2]

12 State two changes to polar areas that are expected as a result of climate change. [2]

Answers online

How to adjust to changing climate

As climates change in extreme environments, people will have to adapt to the changes. This could involve using **intermediate technology** or more sophisticated technology.

Hot arid regions

- Small earth dams can be built by local people to trap and store seasonal rains. They are a cheap and easy form of intermediate technology to build, but thousands would be needed in an area like the Sahel, and they need to be maintained each year.

- **Conservation farming** is a form of multi-cropping that helps conserve water, increase crop **yields** and help resist drought. Up to ten times the usual amount of food can be produced that way, but **NGOs** and farmers need to 'spread the word' across huge areas to make a difference.

- Plant breeding could create new crop varieties that are drought tolerant and can cope with

higher levels of salt in irrigation water. They could be genetically modified, but this is expensive and the research would need to take place in developed countries.

Polar regions

- In Shishmaref, an island off Alaska, coastal erosion due to **sea-level rise** and increased storms is now so severe that the community has decided to relocate to the mainland, but this extreme adaptation risks eroding their traditional culture and lifestyle.

- In Greenland, Arctic cruise ship tourism is increasingly providing people with new income sources, but more tourists could mean more damage to the environment through litter, air pollution and new buildings.

- As permafrost melts, road transport will become more difficult as roads subside and buckle. Air transport is an alternative, but it is very polluting.

The role of global actions in protecting extreme environments

Local actions are a good solution to help people adjust to and cope with a changing environment. Action is also needed globally to try to reduce the threat of climate change. Without these global actions, global warming would eventually over-run local actions. There have been two main attempts at an agreement to reduce **greenhouse gas** emissions:

- The 1997 Kyoto summit tried to reach an agreement to cut greenhouse gas emissions. Unfortunately the USA didn't sign, and though China and India signed, they were not assigned targets.

- The Copenhagen conference of 2009 failed to reach a binding agreement. The so-called Copenhagen accord 'recognises' the scientific case for limiting global temperature rises to no more than 2°C but does not contain any promises to reduce emissions.

There have been international efforts to protect the polar Arctic and areas suffering from desertification:

- The Arctic Council, an intergovernmental group made up of the eight Arctic nations (Canada, Denmark/Greenland/Faroe Islands, Finland, Iceland, Norway, Russia, Sweden, and the USA) and six Indigenous Peoples organisations, was set up in 1996. The Arctic Council has only managed to put forward non-binding recommendations with no enforcement.

- The 1994 United Nations Convention to Combat Desertification aimed to raise the issue of desertification and help countries suffering from it by sharing best practice and funding research. So far it has had little success in halting the spread of deserts around the world.

Exam practice

13 Identify two strategies that could help cope with climate change in polar areas. [2]

14 Describe the role of intermediate technology in improving life in hot arid areas. [4]

15 Explain the role of global actions in helping to protect both polar and hot arid extreme environments. [8]

16 Suggest two reasons why global actions are difficult to implement. [2]

exam tip

If asked for 'actions' in a six or eight-mark question it would be sensible to offer three different ways of coping with climate change with some local detail if possible.

Answers online

Chapter 9 Population Dynamics

How and why is population changing in different parts of the world?

The past, present and future of global population — Revised

- The world population was under a billion in 1750 and had only climbed to 2 billion by 1930. By 2011, it had grown to nearly 7 billion.
- The UN predicts that world population will be between 10.6 and 8.1 billion by 2050 (Figure 1), with 9.3 billion the most likely number.
- The projection range of 2.5 billion is because we cannot know for sure how **birth rates** and **death rates** will change in the future.
- The demands on resources from a world population of 10.6 billion in 2050 would be much higher than a world population of 8.1 billion.

Figure 1 Global population change and projections to 2100

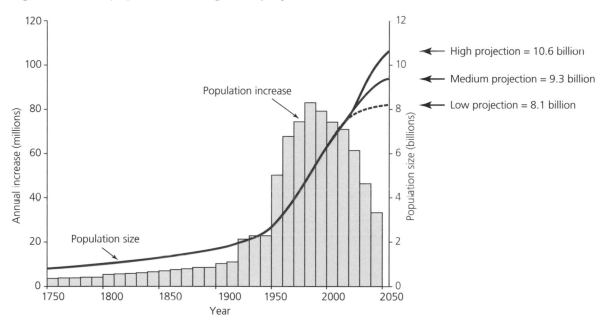

Stretch and challenge

Don't forget that the population is still growing even though the rate of growth of population might slow down.

Check your understanding — Tested

1 When did the global population begin increasing rapidly?
2 Is global population growth getting faster or getting slower?

Exam practice — Tested

1 Study Figure 1. Describe the changes to global population from 1800–2010. [2]
2 Describe the changes in the rate of population growth in Figure 1. [3]

Answers online

Why does population change?

Populations change because of change in birth rates and death (mortality) rates.

exam tip

The study of population is known as **demography**. So if a question asks for 'demographic' changes, these are changes in population.

- Birth rate is the number of children born per 1000 of the population in a year.
- Death rate is the number of people who die per 1000 of the population in a year.
- The **natural increase** of a population is the difference between the birth and death rates. So if the birth rate is 40 per 1000 and the death rate 20 per 1000, the population is increasing at 20 per 1000 or two per cent every year.
- **Life expectancy**: the average age at which people die in a population.
- **Infant mortality**: the number of children under the age of one year who die per 1000 births in a year.
- **Fertility rate**: the number of children that women have in their lifetime. If women have two children or more, the children 'replace' their parents. Fewer than two children means that the population will eventually fall.

The demographic transition model shows how changing birth and death rates affect the total population of a country (Figure 2). Countries move from stage 1 to 5 as they develop economically and socially. No countries are now in stage 1.

Figure 2 The Demographic Transition Model

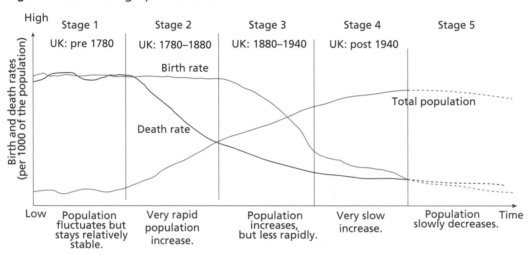

Stage 1	Stage 2	Stage 3	Stage 4	Stage 5
Stable population	Rapidly growing (Uganda, Ethiopia)	Growth begins to slow (Malaysia, Egypt)	Slow population growth (UK, USA)	Declining population (Japan, Germany)
Poor healthcare means a high death rate, but birth rates are equally high	Improvements in healthcare mean death rate falls but birth rate remains high	Social and economic changes mean birth rate begins to fall	Birth rate and death rate balance, so population is stable	Low fertility and very high life expectancy, so birth rate falls below death rate

There are many reasons for the changes between stages 1 and 5 but the key ones are:

- In stage 2, there are rapid improvements in health because of vaccination, medicines and hospitals so death rates fall.
- In stage 3 people are getting wealthier, so do not need as many children to help with farm work. Women also often enter the workforce, so want to have fewer children.
- In stages 4 and 5 birth rates fall below replacement level (two children per couple) because very wealthy societies have few children, so population declines.

Check your understanding

1 Define the term 'natural increase'.

2 Suggest ONE possible reason why natural increase might rise in a country.

Exam practice

3 What happens to birth and death rates in stage 3 of the demographic transition model? [2]

4 Explain why population begins to fall in stage 5 of the demographic transition model. [4]

Answers online

Comparing countries

Many differences in population changes can be seen around the world. In some countries the population is growing rapidly; others suffer from population loss. This is because of:

- differences in birth rates and death rates
- differences in the numbers of people migrating in and out of the country.

Two countries at very different levels of development are Nigeria and Japan (Figure 3). In Japan, population is falling whereas in Nigeria it is rising rapidly. The two countries have very different **population structures** as shown by their population pyramids (Figure 4).

- Nigeria's population pyramid has a very wide base, indicating a youthful population.
- Birth rates are high, but because life expectancy is low, death rates are also high.

- In Japan, the base of the pyramid is very narrow, indicating few children being born.
- Japan has a 'top heavy' pyramid with many people of retirement age – an ageing population.

There are several factors that influence population structure:

1 Economic growth

In Japan, children are seen as a 'cost'. This reduces the number born. In Nigeria, children can be an 'asset' – they can help on the farm or get a street job to boost family income.

2 Migration

Migration can boost a country's population. In the UK, immigration increased population by 2.4 million during the period 1990–2010. In Japan, immigration is very low so does not help offset its ageing population. About 300,000 people emigrated from Nigeria in 2012 – mostly working-age men – to seek better jobs in other countries.

3 Conflict

War and conflict often lead to a lower number of men in a country. This can be seen in the pyramid for Japan, as the number of men over 80 is very small – an impact of the deaths that occurred in the Second World War.

Figure 3 Nigeria and Japan compared

2012 data	Nigeria	Japan
Total population	162 million	128 million
Birth rate	40 per 1000	9 per 1000
Death rate	15 per 1000	10 per 1000
Population growth rate	2.5%	–0.1%
Life expectancy	51	83
Per capita income (US$)	1,500	47,000

Figure 4 Population pyramids for Nigeria and Japan

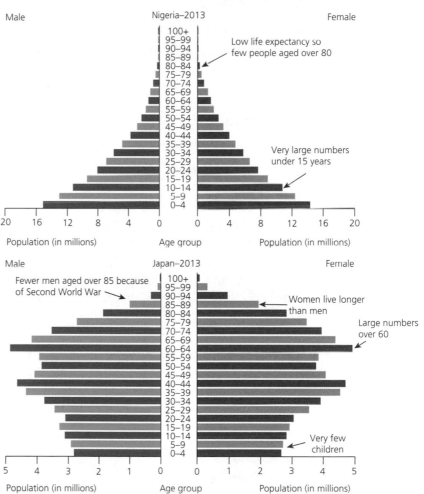

Check your understanding

1 State TWO pieces of evidence that Nigeria is a developing country.

2 Suggest TWO reasons why developed countries have lower death rates.

Exam practice

6 Compare the population pyramids of Japan and Nigeria. [4]

7 What evidence is there in Japan's population pyramid that is has an ageing population? [4]

Answers online

Population structures

Some countries, like Japan, are ageing rapidly. This is because of falling birth rates, very good healthcare and so very high life expectancy, and low birth rates due to the costs of having children. In many developed countries, women have established careers and so delay having children into their 30s to focus on their jobs. High childcare costs are also a factor, plus the high cost of housing means people delay marriage. Ageing populations can bring a number of challenges:

● Healthcare costs and demands rise, and in the UK this can put pressure on the NHS.

● The cost of long-term residential care and nursing homes for the elderly is rising.

● As the percentage of people aged over 65 rises, the number of working-age people falls – the smaller number of workers will have to pay higher taxes to care for the elderly.

● A shortage of workers for offices and factories is an issue facing Japan today.

Very old people and very young people are called dependents because they do not work. Japan has a high dependency ratio because of its ageing population.

Check your understanding

Tested

1 Identify ONE economic reason why birth rates in Japan have fallen.

2 Identify TWO problems caused by Japan's ageing population.

Knowing the basics

An ageing population is one where the average age is rising.

Nigeria's youthful population

Young populations can also be difficult to manage.

Nigeria has more or less the same **population structure** today that it had in 1975, with 75 per cent under the age of 30. Despite its wealth from oil, most of the population is very poor. The costs of feeding this population and funding its education are high.

The total fertility rate remains extremely high. Only eight per cent of married women of reproductive age use a modern method of contraception, partially because they want large families. Several things explain this desired fertility, including poor child survival – one-fifth of all children born in Nigeria die before five years of age – and low educational attainment among women, 42 per cent of whom have never been to school. In twenty years' time, Nigeria's very young population will be of working age. This could provide a huge boost to the economy in the future.

Knowing the basics

In developing countries such as Nigeria women have limited opportunities – this is a cause of high birth rates.

Stretch and challenge

When women are given basic education they have more choices and fertility rate almost always falls.

Check your understanding

Tested

1 Identify TWO economic costs of a young population.

2 Identify a possible economic benefit of a youthful population.

Exam practice

Tested

8 Using named examples, compare the challenges for countries with youthful and ageing populations. [8]

Answers online

How far can population change and migration be managed sustainably?

Why do countries wish to control their populations?

Revised

In any country there will be a wide variety of opinions about the 'ideal' or optimum population size.

Figure 5 Relationships between population and resources

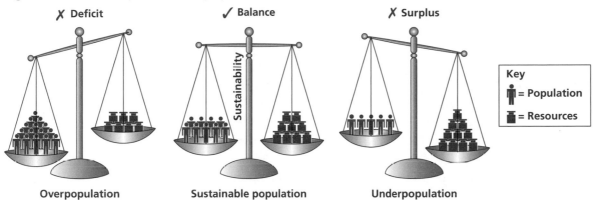

Some governments feel they need to manage their population actively, rather than just let population change happen naturally. There are several reasons for this:

1 Pressure on resources

Rapid population growth could put pressure on food, water and energy supplies. Restricting population growth could relieve this pressure. Most countries can trade to get the resources they need so this reason is less important nowadays.

2 Overcrowding

Resources like space, housing, education and health services could be overstretched by rapid population growth, especially in small countries like the UK or Singapore.

3 Ageing

Countries like Japan could offset the problem of ageing by encouraging a higher birth rate and/or encouraging immigration, but both of these are controversial concepts.

4 Skills shortages

Some countries may lack the skilled workers they need, in which case they often encourage immigration. Increasing the amount of skilled workers by raising birth rates rather than educating children is possible, but would take too long.

Managing population is controversial. The issue of immigration is a 'touchy subject' for many people. Trying to force people to have fewer or more children raises human rights issues.

> **Knowing the basics**
>
> If population grows faster than resources there will be a crisis of some sort. Famine and war are obvious possibilities.

> **Stretch and challenge**
>
> Ageing populations can also 'save' money for governments. They spend a lot less on children's education and maternity care.

> **exam tip**
>
> You will need to know two case studies for this section: one of a country with **anti-natal policies** that discourages large families and one with **pro-natal policies** that does the opposite. You may be asked to provide a few details, so learn three facts about each country.

Check your understanding

1 What is overpopulation?

2 What is underpopulation?

Tested

Exam practice

Tested

9 Define the term 'immigration'. [2]

10 Explain how immigration could help manage skills shortages and ageing in a population. [4]

Answers online

Population policies in practice

China: anti-natalist but for how long?

Figure 6 Trends in China's population growth rate, birth rate and death rate

Why anti-natalist?

- The population was growing very rapidly throughout the 1950s and 1960s – at that time encouraged by the government.
- New political leaders decided that a large population size was a problem.

Which policies?

- Voluntary programmes and land reform led to a sharp fall in birth rate in the 1970s.
- The famous 'one-child policy' was started in 1979. It gave benefits to women in the form of cash bonuses, better housing and maternity care. It also punished couples who did not sign up. Forced sterilisation happened in some regions of the country.

The impact?

- The birth rate continued to fall. Today China has a fertility rate of 1.7; below the replacement level of two.
- This has raised issues of China's rapidly ageing society and the impact on the single children of looking after elderly parents.
- The preference for boys (still common in many societies) led to sex-selective abortion and today there are 120 males for every 100 females.
- A 'social' impact much commented on is how spoilt the single children are – described by some as 'little emperors'.

Singapore: first anti-natalist and then pro-natalist!

- Singapore is a small, island nation. When it first became independent it pursued an anti-natal policy just as in modern China, although not quite so strictly.
- This policy was successful, and along with the impact of later marriage, rising incomes and the increased role of women in society, the fertility rate fell from 3.0 in 1970 to 1.6 in 1985.

- Today Singapore is worried that its only resource is its population so the government now tries to encourage earlier marriage and larger families.
- Couples with three or more children pay lower taxes, have better housing, easier access to nursery schools and preference in school choice later on.
- The policies have had a very limited impact in Singapore with the drive today being to get couples together in the first place. The government is so desperate it now sponsors speed-dating events.

Stretch and challenge

Many countries, such as Singapore, have few resources except their people. Population growth provides more skills if these people can be educated.

Exam practice

11 What is a pro-natalist population policy? [2]

12 Using a named country, explain why it introduced anti-natalist policies and the impact of these. [8]

13 Why did Singapore 'switch' population policy from anti- to pro-natalist? [2]

Answers online

Why do migration policies vary from place to place and from time to time?

Most countries have policies on migration. These either attempt to reduce migration or increase it. Reasons for reducing immigration include:

- the unpopularity of large-scale immigration among voters
- fears that immigrants accept lower pay, reducing pay for everyone
- fears that the host country culture becomes 'swamped' by immigrants and that large-scale immigration can increase cultural tensions.

Most countries work hard to police their borders and prevent illegal immigrants, but are open to genuine refugees and asylum seekers who are fleeing persecution. The real issue is how many legal economic migrants a country should accept.

In some cases countries want to increase immigration to:

- reduce skills shortages and help the economy keep growing
- offset the problem of ageing by attracting working-age immigrants
- attract low-skilled, low-wage workers for farming and construction because the existing population won't do the 'dirty, dangerous and demeaning' jobs.

Knowing the basics

Quotas and skills tests are the commonest ways of restricting numbers of migrants.

Different migration policies and their impact

Evaluating the UK's immigration policy

Most immigration to the UK is from the EU (especially Poland, Ireland, Germany and France). Immigrants also come from the Commonwealth (India, Pakistan, Australia) and from the USA, China and the Philippines. About 60 per cent of immigrants come to work and about 40 per cent to study. Only non-EU immigration can be restricted.

	Arguments for	Arguments against
Open-door policy to Eastern European EU migrants since 2004	Provides many low wage workers in farming, fish processing and low paid services, filling a 'gap' in the labour force	The number who have arrived is much higher than forecast, putting pressure on services like schools and housing
Points-based immigration since 2008 for non-EU migrants	The points system allows skills and migrant type to be matched with the UK's needs (skills tests)	It is a complex system that might deter some valuable migrants, such as entrepreneurs, who go to another country
Cap (quota) on non-EU immigration since 2010	Key to reducing overall immigration numbers because the UK cannot restrict EU immigration	Once the annual quota is filled, no more immigrants are allowed, so some **TNCs** cannot get the skilled workers they need

Knowing the basics

You need to know why migrants come to a country. Work is the main reason – an economic motive. Education and joining relatives are common social reasons.

Stretch and challenge

All governments know that they need some migrants to fill jobs that cannot be done as well as by the 'locals' and most know that in a globalised world there may have to be more migration in the future in order to stay competitive.

Check your understanding

1 Name TWO global locations that supply most of the migrants to the UK.
2 Identify the TWO main reasons why migrants come to the UK.

Exam practice

14 Outline two reasons why countries may wish to encourage immigration. [4]
15 What is an immigration 'quota'? [1]
16 Using a named example, explain how one country has attempted to manage migration. [6]

Answers online

Chapter 10 Consuming Resources

How and why does resource consumption vary in different parts of the world?

Resources can be classified in a number of ways. There are **natural resources** (found in the physical environment around us) and **human resources** (people and their skills). Natural resources are classified by type:

Mineral	Physical	Biological	Energy
Ores like iron ore and bauxite (for aluminium), gold, diamonds and rock salt	Water, the sun's energy, the wind and the land (as well as its soil)	Anything living, such as forests, fish in rivers and oceans, and animals	Fossil fuels (coal, oil, gas) which are the buried remains of ancient plants and animals

Many types of resources can be used as a source of energy (heat or power), not just fossil fuels. For instance, biological resources (wood) can be burnt to release energy. Water, the sun's energy and wind can all be harnessed to generate electricity (**HEP**), solar panels and wind turbines).

Type of resource	Definition	Examples
Non-renewable	There is a finite 'stock' of these, usually in the ground. Once humans use the stock up, no more will be available	All fossil fuels and ores/minerals
Renewable	There is a continuous 'flow' of these resources (really a flow of energy), which is infinite so will never run out	Physical resources used for renewable energy such as wind, solar and wave power
Sustainable	These resources can be used but then replaced, if humans manage them carefully – most are biological	Timber from forests can be cut and replanted. Fish can be repeatedly harvested if we don't take too many

Remember that we are considering resources on a human timescale. Fossil fuels take millions of years to form, so even though some are forming today, they will not be 'ready to use' on an appropriate timescale.

exam tip

Make sure you can classify a range of natural resources, and use words such as 'finite' and 'infinite' carefully.

Exam practice ———————————————————————————— Tested

1 Define the term 'renewable resource'. [2]

2 What are biological resources? [2]

Answers online

Resources are unevenly distributed and unevenly consumed

A non-renewable energy resource: oil

Many countries have oil reserves. This is really an 'accident' of geology – the oil-bearing rocks happen to be under certain countries. How much oil a country produces also depends on:

- having the money and technology to develop oil wells; one 'dry' well can cost US$200 to drill
- political decisions to protect areas from oil drilling, such as the USA's Arctic National Wildlife Refuge, or develop tar-sands in sensitive areas, such as the boreal forests of Athabasca in Canada.

Figure 1 shows that oil production grew between 1970 and 2010, and is expected to increase more by 2030. By 2030:

- the Middle East could be supplying close to 50 per cent of world oil, because it has the most remaining oil reserves
- Asia, Europe and Russia's oil production will be less important because oil reserves will be running out
- North America's production is expected to rise because of developing new, controversial oil sources like tar-sands, oil-shales, and drilling in the Arctic and far out to sea.

Figure 1 Changing oil production

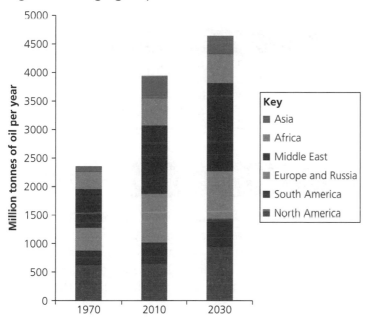

Knowing the basics

Wealthy, developed countries consume much more oil because they have more cars, more consumer goods and more expensive lifestyles.

Stretch and challenge

Some countries consume a great deal of oil because they produce it very cheaply – people may not all be wealthy but petrol is cheap. Saudis can buy 15 litres for the same price as 1 litre in the UK.

Oil **consumption** has dramatically changed (Figure 2). In 1970 Europe and North America consumed 76 per cent of all oil. By 2030 this will drop to 37 per cent. In Asia the percentage will rise from 15 per cent to 39 per cent. The changes are driven by:

- economic growth and rising incomes in the developing world, especially in China, India and Brazil
- demand from industry for fuel 'made in China'
- using oil more efficiently in the developed world, so consumption stays fairly static.

Exam practice

3 Using Figure 1, describe how oil production changes between 1970 and 2030. [4]

4 Give two reasons why some regions produce more oil than others. [2]

Answers online

Figure 2 Changing oil consumption

Oil consumption (millions of tonnes of oil per year)	1970	2010	2030	% change 1970–2030
Africa	35	161	228	+550
Middle East	59	364	527	+790
South America	104	281	395	+280
Asia	338	1281	1859	+450
North America	798	1041	926	+16
Europe and Russia	928	903	824	–11

A renewable energy source: hydroelectric power (HEP)

HEP is a form of electricity generation. A dam is built across a **river valley**, creating a **reservoir**. Water from the reservoir flows down pipes in the dam and the force of the water turns turbines which generate electricity. To develop HEP, you need:

● valleys, usually in mountains, to build dams across

● high rainfall and large reliable rivers

● money and technology to build and install the dam, turbines and electricity pylons.

HEP is produced in one country and then generally used in that country. This is because electricity is not traded globally (unlike oil).

HEP is not found in:

● desert and low rainfall regions, or areas with very seasonal river flows

● poor developing countries which lack the money to build dams and develop complex electricity grids.

The future pressures on supply and consumption — Revised

Both oil and HEP use are under pressure. Demand for oil is constantly rising as people around the world enjoy economic growth:

● The Chinese, Russians, Indians and Brazilians hope to enjoy the same income and lifestyles as Europeans and Americans.

● 25 per cent of global oil is consumed in the USA by only 5 per cent of the world's population.

● Oil is used to produce petrol, diesel, jet fuel, most plastics and paints – demand is rising.

The amount of oil is finite, which leads some people to believe we will soon reach 'peak oil' – the point at which we can increase production no further. This could lead to increasing political tension as consumer countries try to negotiate deals with producing countries. China has built alliances with many African oil producers such as Sudan. Resource conflicts are a distinct possibility with consumption rising and production static or falling. Countries may decide to keep their own oil and not trade it.

HEP may seem like a more stable energy source, but could soon be under threat:

● Building dams in one country changes river flow, so countries downriver object – this has happened on the Colorado River between the USA and Mexico.

● **Climate change** could reduce water supplies, so rivers run dry and electricity stops.

● Big dams, like China's Three Gorges, are controversial and upset local people.

Exam practice — Tested

5 Using Figure 2, describe how oil consumption changes between 1970 and 2030. [4]

6 State one reason why non-renewable resources cannot be produced in all countries. [1]

7 Explain why oil consumption is likely to increase in the future. [4]

Answers online

How sustainable is the current pattern of resource supply and consumption?

Theories about population and resources

Malthus (1766–1834)

- It seems obvious that as the population grows the fewer resources we will have. Thomas Malthus concluded this over 200 years ago in his **basic theory**.
- He suggested that population would grow by doubling: 2, 4, 8, 16, 32 and so on, but food production would only increase singly: 2, 3, 4, 5, 6 and so on.
- Obviously a gap would appear between the two and food shortages would lead to a series of social and economic crises with an eventual collapse of the population.

Figure 3 A graph showing Malthus' basic theory

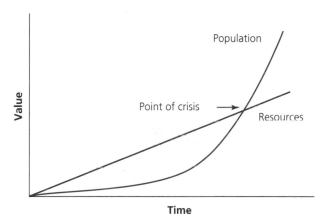

Knowing the basics

When population is larger than the resources available there will be conflict, disease and famine.

Stretch and challenge

Modern Malthusians are less troubled by population growth but more worried about the uneven pattern of consumption with rich countries consuming so much.

Boserup (1910–1999)

- Ester Boserup took a very different view to Malthus. The rapid growth of the human population in the past two centuries does pose a challenge for Malthusians. **Boserup's theory** suggested that we never run out of resources because as we get to the point when resources are getting short we are pressured to invent ways of avoiding a crisis.
- For Boserup, a growing population *causes* changes in technology that allow the population to grow again. This makes population growth absolutely central to the development of the human species.

Figure 4 A graph showing Boserup's theory

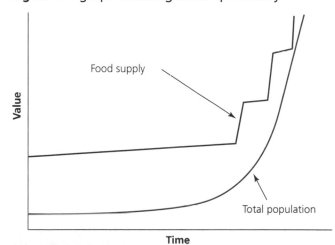

Knowing the basics

'Necessity is the mother of invention.' Population growth leads to changes in technology allowing the population to grow again.

Stretch and challenge

Boserup believed that population grows because we make changes in technology. In her view, the 'green revolution' and the development of GM crops are *results* of population growth.

Exam practice

8 Explain the view of Malthus on the relationship between population and food supply. [4]

9 According to Boserup, when population rises what happens to food supply? [2]

Answers online

Global food supply and demand

The question of who is right, Malthus or Boserup, is difficult to answer. If Malthus was correct, there should be evidence of food shortages (famine) because the population has outstripped food supply. There have been famines such as in Ethiopia 1984–85 and North Korea 1994–98, but these only affected small areas, and were temporary. In the two centuries since Malthus developed his theory the population has risen from 1 billion to over 7 billion, and most of those 7 billion are reasonably well-fed.

If Boserup was correct, food production should have risen continuously as better farming technology has been developed. This is what has happened in most regions (Figure 5):

● Farm machinery was introduced in the late nineteenth century, along with artificial fertilisers.
● After the Second World War, pesticides and herbicides were developed to increase **yields** by killing pests, diseases and weeds.
● In the 1960s the 'green revolution' developed high-yield varieties of maize, wheat and rice by plant breeding.
● Since the 1990s the 'gene revolution' has developed new genetically modified (GM) crops to resist disease and drought.

Knowing the basics

Malthus was proved wrong by the growth in food output and wealth in the nineteenth and twentieth centuries.

Figure 5 Food production per person 1961–2011

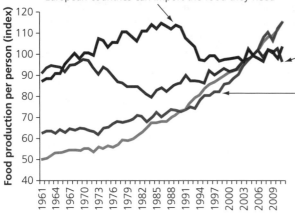

In Europe, food production fell in the 1990s and is now stable – but rich European countries can import the food they need

In Africa, there has been much less progress – food production per person was very similar in 2011 compared to 1961

In Asia and South America food production per person has more than doubled since 1961, despite population growth

Key
— Europe — Asia
— Africa — South America

Stretch and challenge

In reality Malthus was only concerned about the growth of the 'poor' population. He was worried about a revolution.

However, we should not forget about Malthus:

● There are still about 850 million malnourished people on the planet, so not everyone has enough food.
● Rising food prices can quickly deprive people of enough food. World average food prices rose 70 per cent from 2001 to 2011.
● Just because the world can feed almost 7 billion people now does not mean it could feed 9 or 10 billion in 2050.
● Farming depends on water and oil, both of which may soon become scarce.
● Problems like soil **erosion** and **desertification** could rapidly worsen because of **global warming**, reducing food production.

Exam practice

10 Using examples, consider how far Malthus and Boserup were right. [8]
11 Using Figure 5, describe the trends in per capita food production for Africa and Asia. [4]

Answers online

Resource consumption needs to be made more sustainable. This really means:

- reducing resource consumption per person, so **non-renewable resources** last longer
- reducing food, water and energy waste, so resources are used more efficiently
- using more environmentally-friendly resources with lower carbon emissions to slow global warming.

It could also mean using resources in a 'fairer' way. People in the poorest, hungriest parts of the developing world actually need more resources – perhaps this could be achieved by people in the richest, greediest parts of the developed world using less?

There are two ways of reducing resource consumption:

- Reducing our consumption both as individuals and as countries.
- Making products more efficiently by using more sustainable materials and with less waste.

Changing our lifestyles (individual)

- Take public transport and give up using your own car.
- Eat only vegetarian food.
- Use low-energy light bulbs.
- Turn down the thermostat on heating.
- Reduce water use to a minimum.

Changing production methods

Interface Carpets is a company trying to reduce its impact on the environment. The company will:

- obtain all of its energy from renewable sources by 2020
- install renewable energy systems
- measure, reduce and compensate for carbon emissions
- sell 'carbon neutral' products
- motivate workers with waste reduction programmes.

Both local and national government can help achieve a more sustainable resource future. National governments tend to do this using laws and targets.
Local government often does this in more practical, 'hands-on' ways:

Education	**National government:** the UK national curriculum includes sustainability, recycling and environmental issues as part of science and geography **National government:** Recycle Now (www.recyclenow.com) is the UK's national campaign to educate people about the need to recycle and provides advice on how to do it
Conservation	**National government:** has introduced variable car tax, to encourage people to buy smaller, more efficient cars as well as phasing out incandescent light bulbs to encourage people to buy energy efficient bulbs **National government:** has set a target that 15% of electricity generation should be renewable by 2020 **Local government:** provides and manages green spaces such as parks, **greenbelts**, local nature reserves
Recycling	**National government:** sets targets for recycling. Recycling household waste in the UK increased from 11% in 2001 to 40% in 2011 **Local government:** manages household waste, by providing home recycling bins as well as recycling plastics, food waste, paper, glass and other materials

Exam practice — Tested ☐

12 What is meant by 'sustainable resource consumption'? [3]

13 Explain how both national and local government can attempt to manage resource consumption. [6]

14 State two ways individuals could reduce their resource consumption. [2]

Answers online

The prospects for a switch to alternative and renewable resources

Revised

The full definition of sustainability is taken from the United Nations report of 1987:

Sustainable development is development that meets the needs of the present without compromising (limiting) the ability of future generations to meet their own needs.

Knowing the basics

We should consume in such a way as to allow our children and grandchildren the chance to enjoy a decent lifestyle.

Stretch and challenge

Many people lack even the most basic resources such as food, fuel and water. They are likely to be much more concerned about gaining any resources they can, rather than worrying whether or not the resources they use are sustainable.

There are a number of ways resource shortages could be tackled:

Problem	Solutions	Prospects and Challenges
Fossil fuels are a finite resource	• Hydrogen could be used as a fuel to replace petrol and diesel (from oil) • When hydrogen is burnt it only releases water vapour, not **carbon dioxide (CO_2)** • Maize and other crops can be made into biodiesel and bioethanol as renewable alternatives to petrol and diesel	• Hydrogen fuel cell technology works, but it is very expensive to make, store and safely transfer hydrogen. It is a very explosive gas • Biofuel crops use land that once grew food, so increasing biofuels could reduce food supply and/or increase food prices
Food supply is under pressure as population rises	• GM crops could be developed that are drought or flood tolerant, resistant to diseases and have higher yields • Drip irrigation and zero-tillage farming are both methods of using less water on farms by limiting **evaporation**	• The public, at least in the UK, are uncomfortable with GM as it is seen as 'meddling with nature' • Farmers would need to invest in expensive new equipment and change how they farm, which would take time
Many resources are simply wasted	• Bottled water could be banned – it is very wasteful (for example, the bottles are made out of plastic bottles, and we use fuel to transport it) and many bottles end up as litter • People could be encouraged to compost food waste in their gardens, or recycle all food waste – this compost could be a renewable alternative to chemical fertiliser	• Many people would see this as an infringement of their rights, although bottled water was banned in the Grand Canyon in 2012 due to litter • Some people do not have gardens; council-run schemes need new infrastructures and bins, and are expensive to set up

Exam practice

Tested

15 State two alternatives to using fossil fuels to power motor vehicles. [2]

16 Using examples, consider ways in which new technology could help combat resource shortages. [8]

Answers online

Chapter 11 Globalisation

How does the economy of the globalised world function in different places?

Globalisation Revised

Globalisation means the greater integration of the **global economy**. It is a process that has accelerated dramatically over the last 30 years. It means:

- jobs and wealth in one country are increasingly dependent on trade and investment from other countries
- countries specialise in certain types of economic activity, like finance in the UK or manufacturing in China

- jobs move from high cost countries to low cost countries
- the amount of trade in **goods**, services and money is increasing
- the world is increasingly connected by air travel, shipping routes, the internet and other forms of communication.

Changing employment structure Revised

Despite globalisation and increasing wealth in the world, countries are at very different stages of **economic development**, with different employment sectors dominating. The **Clark–Fisher model** illustrates this (Figure 1).

In post-industrial economies, secondary industry declines and a new sector emerges – the **quaternary sector**. It includes scientific research such as biotechnology and computer science, as well as hi-tech manufacturing, for example computer chips. Few people are employed in this sector but it is very valuable.

Stage	Dominant economic sector	Examples (% employment Primary/Secondary/Tertiary)
Pre-industrial	**Primary:** farming, fishing, mining and forestry which grow or extract raw materials	Ethiopia (85/5/10) Nepal (75/7/18)
Industrial	**Secondary:** manufacturing goods in factories and workshops	China (38/29/35) Mexico (14/23/63)
Post-industrial	**Tertiary:** services such as education, retail, banking, the health service and travel	UK (2/18/80) Japan (4/26/70)

Figure 1 The Clark–Fisher Model

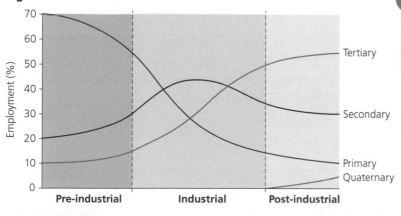

exam tip

All topics have key terms to learn. In this topic, these include 'globalisation' and the names of the four economic sectors. Make sure you use these in your exam answers.

Check your understanding

Give examples of types of jobs in the four employment sectors.

Tested

Exam practice Tested

1 Explain what is meant by 'globalisation'. [2]

2 Describe how employment changes as countries undergo economic development. [4]

Answers online

Contrasting employment sectors

As countries develop, and move through the stages of the Clark–Fisher model, employment changes from primary to secondary to tertiary. This brings other changes such as:

● an increase in pay and therefore income
● better working conditions
● a change from informal to formal jobs.

In developing countries many people work in an informal job, for example a street seller or subsistence farmer. These jobs are unregulated, untaxed and the work is often difficult and dangerous. Formal jobs have contracts, regular (and taxed) pay and are regulated, for example, with health and safety laws and holidays. Working conditions can be very different across the sectors:

Job and sector	Working conditions	Pay
A subsistence farmer in Ethiopia (primary)	Hard, labour intensive farm work Vulnerable to the **weather**, e.g. floods and drought Workers' lives depend on producing enough food Children in the family often have to work	No pay. Might earn a small amount selling any surplus food produced
A factory worker in China (secondary)	Long hours, repetitive work, lots of overtime Unions are often banned so employees have fewer opportunities to exercise their rights Most factories only want workers under 30 years of age.	£1000–£3000 per year
A nurse in the NHS (a government organisation) in the UK (tertiary)	Good working conditions with health and safety regulations and union representation Paid holidays and pensions Unsociable working hours and a high stress working environment	£20,000–£40,000 per year
A pharmaceutical researcher in the USA (quaternary)	Very good working conditions High stress job with TNC demand to develop new products	£25,000–£100,000 per year

Global institutions and globalisation

Globalisation has resulted in some key changes to employment sectors in the last 30 years:

● In the developed world, secondary (manufacturing) jobs have been lost, but industrialising countries in the developing world, such as China, have gained secondary jobs.

● In industrialising countries the number of people working in primary jobs has fallen – many people have moved to cities to work in factories.

● Even some tertiary jobs, like those in **call centres**, have moved from developed countries like the UK to developing countries such as India.

These major changes in the location of jobs are often referred to as the **global shift**.

What has caused these changes? Three global institutions are important as they have helped create a more globalised world economy:

Global institution	Role in globalisation
World Trade Organisation (WTO)	Promotes free trade by persuading countries to reduce or remove trade barriers like taxes, tariffs and quotas
International Monetary Fund (IMF)	Gives loans to developing countries for infrastructure and encourages countries to allow foreign investment to create new jobs
Transnational Corporations (TNCs)	Aim to reduce costs and increase profits by moving factories to cheaper locations. This creates new jobs in developing countries

Exam practice

3 What are 'informal' jobs in the developing world? [2]

4 Using examples, compare the working conditions of people in the developed and developing world. [6]

5 How has the location of secondary employment changed in the last 30 years? [2]

Answers online

TNCs are perhaps the most important type of institution. They are such huge, global companies that they can influence whole economies. McDonald's has 34,000 restaurants in 119 countries which employ 1.8 million people.

The impact of globalisation on different groups of people

Revised

Globalisation has had an impact on many different groups of people. In the last 30 years:

- most countries have become richer, especially in the developed world
- about 300 million people in China have been lifted out of poverty
- elsewhere, especially in Africa, incomes have not improved much at all.

We can look more specifically at the impact of globalisation on different groups of people in the table below. Some groups have benefited from new jobs although these jobs may be low paid with poor working conditions:

Male car factory workers in the USA	Female factory workers in China
Companies like Ford and General Motors have shut car factories in cities like Detroit in the USA, and moved them to Mexico and Brazil. Hundreds of thousands of male, well-paid workers lost their jobs as the jobs moved overseas to cheaper locations	Women who work for Foxconn, the company that makes iPads and iPhones for Apple, earn about £180 per month. Most live in dormitories within the factory and work up to 60 hours of overtime a month, but pay is much higher than in the rural areas from which most workers migrated
Male coal miners in China	**Female call centre workers in the UK**
Coal has powered China's economic growth, providing jobs for 5 million coal miners who mined 3500 million tonnes in 2011, though mining is dangerous. Up to 2000 miners die each year working 7-hour shifts underground for £5–£8 a day	About 900,000 people in the UK work in call centres. Many are part-time and most are women. They earn around £15,000 a year. This industry has grown in the last 20 years, and has provided women with jobs, although it is low-skilled work and promotion prospects are limited
Male TNC executives	**Female textile workers in Bangladesh**
Most of the top jobs at TNCs are taken by men, with women making up 18% of senior managers. TNCs have grown because of globalisation and so have the executives' wages; since 1980, their wages have increased by 3000%	Many low-priced clothes are made in Bangladesh by women working for up to 80 hours a week but only earning as little as £12. Accidents, factory fires and exploitation are common – all in the name of cheap clothing

Exam practice

Tested

6 Describe how global institutions have helped create a more globalised economy. [4]

7 Identify one group of people who have lost their jobs because of globalisation. [2]

8 Using examples, explain how globalisation has had both positive and negative impacts on different groups of people. [8]

Answers online

Check your understanding

Tested

Refer to the table above. How different are wages in different jobs around the world?

What changes have taken place in the flow of goods and capital?

International trade and FDI

Money flows around the world in several ways:

● Through trade in goods (imports and exports) such as manufactured products, oil, food and raw materials

● Through stock markets

● When money (**capital**) from one country is invested in another, for instance, to build a new factory. This is called **foreign direct investment (FDI)**.

The volume of these movements has increased. In 1960 all of the world's exports were worth US$0.3 trillion. This reached US$3.3 trillion in 1990 and US$15 trillion in 2010. The exporters have also changed (see Figure 2):

Figure 2 World exports 1970–2010

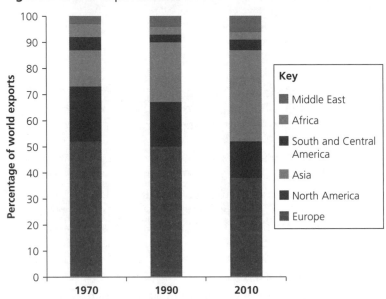

● Africa, the Middle East, and South and Central America account for about the same percentage of exports in 2010 as in 1970.

● Asia's share has grown from 14% to 35% because countries like China, India and South Korea have industrialised.

● In the developed world, Europe and North America, the percentage of exports has fallen.

China's share of world trade (all imports and exports) has grown from less than 1 per cent in 1970 to over 11 per cent in 2010. The USA's share declined from 16 per cent in 1970 to only 8 per cent in 2010.

Figure 3 shows changes in types of trade:

Figure 3 World exports in 1990 and 2010

World exports by product (%)	1990	2010	% change 1990–2010
Fuels	11	18	5
Chemicals	9	12	3
Ores and minerals	4	5	1
Iron and steel	3	3	–
Electronics and ICT	9	9	–
Cars and trucks	9	7	–2
Food and farm products	12	9	–3
Clothing and textiles	6	4	–2
Other manufactured goods	37	33	–4

The share of world trade has increased for fuels like oil, coal and gas, and raw materials such as chemicals, ores and minerals. This is because industrial countries like China and Brazil need huge quantities of energy and raw materials for their factories.

Another major change has been the amount of FDI. Usually, a TNC will invest money (capital) from one country into another. This could be to open a mine, build a factory or finance a dam. Seventy-five per cent of FDI capital originates in developed countries.

Figure 4 FDI 1970–2010

Billions of $ of FDI per year (USD)	1970	1980	1990	2000	2010
Developed countries	9	47	172	1137	618
Developing countries	4	7	31	214	501
China	0	0.1	3	41	114

Check your understanding

Which types of countries receive the most FDI?

Tested

Figure 4 shows that:

- most FDI goes to developed countries (and always has), for example, the British TNC British Petroleum (BP) investing in America
- the amount of FDI going to developing countries has increased, for example, a UK company opening a call centre in India
- the amount of FDI going into China has grown from almost US$0 in 1980 to US$114 billion in 2010.

FDI is important because investment creates jobs. China's large FDI explains why it is now the 'workshop of the world'.

Knowing the basics

Trade has grown hugely over the last 50 years and this has helped many people out of poverty.

Stretch and challenge

FDI can be a double-edged sword; a TNC investing in China may be doing this by removing investment (closing a factory) somewhere else like the UK.

exam tip

Questions which say 'using examples' require more than just naming a country or continent; try to include specific locations, facts and details.

Exam practice

Tested

9 What is FDI? [2]

10 Using examples, describe how the volume and pattern of world trade has changed over the last 50 years. [6]

Answers online

The drivers of change

Revised

It is important to realise that trade and FDI are much easier to implement than in 1960. Back then, it was difficult to keep track of foreign investments and it might have taken months to get goods or raw materials from Asia. Things have changed:

● Communication is cheap and instant because of fibre optic internet cables, satellites and technology such as computers and mobiles phones.

● Jet aircraft have reduced the cost of travel and connected up distant places.

● Container ships have revolutionised trade in goods, making it cheap and efficient to move products from producer to consumer.

One container ship can hold 12,000 pairs of trainers, and costs about £3000 to ship from China to the UK. Shipping costs only add about 25p to the cost of each pair of trainers.

The rise of TNCs has also helped globalise the world economy:

● TNCs are large, wealthy, powerful and can transfer investment around the world.

● They try to maximise profits by reducing costs, and often move production to low cost places such as Asia.

● TNCs have complex networks of factories, offices, HQs and suppliers, as well as expertise in managing these global networks.

● The number of TNCs has risen from about 7000 in 1970 to over 80,000 today.

Mergers have made some TNCs even larger and more powerful. One of the biggest, Exxon-Mobil, formed in 1999 as a merger between Exxon and Mobil. In 2011 it had sales of $486 billion, almost as much as Belgium's GDP for that year.

State-led investment has also encouraged trade and FDI in two main ways. First:

● In China, India and other Asian countries governments have set up **free-trade zones (FTZs)** and **special economic zones (SEZs)**.

● TNCs can build factories and offices in these purpose-built industrial estates cheaply.

● Unions are usually banned, taxes are low and regulations on **pollution** and health and safety are minimal.

Second, many large companies are owned and controlled by governments as 'state-owned enterprises'. This is especially true in the developing world. China has three large state-owned oil companies: Sinopec, CNOOC and CNPC. To secure China's oil supply, these companies have undertaken FDI in Indonesia, Burma, Sudan, Gabon, Thailand and even Canada.

Stretch and challenge

Some people are concerned about the power of TNCs. Certain TNCs are now almost as powerful as some national governments.

Check your understanding

Tested

How do governments in Asia encourage TNCs to invest?

Knowing the basics

TNCs can bring prosperity and jobs to areas they invest in.

exam tip

You need to know the difference between 'describe' (what) and 'explain' (why). 'Describe the trends in Figure 4' is a very different question to 'Explain the trends in Figure 4'.

Exam practice

Tested

11 Describe how transport and communication have helped the world globalise. [4]

12 What is state-led investment? [3]

Answers online

TNCs operate in at least two countries. Volkswagen Group (VW) is a TNC with its HQ in Wolfsburg, Germany. It produces cars, vans, trucks, buses and motorcycles. It is a large global company:

● it has 94 production sites in 24 countries

● it employs over 500,000 people

● it sold 9.2 million vehicles in 2012, worth £160 billion.

VW has expanded its production (Figure 5) to become a worldwide company, the third largest car company in the world.

Figure 5 VW's global production network

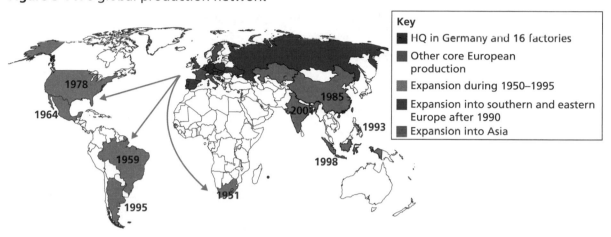

One of the major ways VW has grown is by merger. Over decades, it has absorbed the companies Audi, SEAT, Skoda, Bentley, Bugatti, Lamborghini and Porsche. It has also shifted production overseas to cheaper locations with lower wages – first to South and Central America, and more recently to eastern Europe and Asia. This has altered VW's balance of production in Germany versus its production abroad:

	1990	2000	2012
Production in Germany	1.8	1.9	2.3
Production abroad	1.2	3.2	6.9
Employees in Germany	166,000	167,000	237,000
Employees abroad	95,000	157,000	296,000

These figures show the global shift. In 1990, 64 per cent of VW workers were in Germany but by 2012 German workers made up only 44 per cent. Vehicle production was 60 per cent in Germany in 1990 but 75 per cent abroad by 2012. Smaller 'German'-branded VW cars have been **outsourced** to production plants abroad. The VW Up is produced in the Czech Republic and the VW Polo in Spain. The wages are lower than in Germany which is important as small cars have small profit margins.

Check your understanding

Tested

How does VW illustrate the 'global shift'?

Stretch and challenge

VW provides many skilled, relatively well-paid jobs worldwide, although its parts suppliers may not have such high standards.

Exam practice

Tested

13 What is a TNC? [2]

14 Explain why many TNCs have shifted production to Asia. [4]

Answers online

A tertiary sector TNC

Revised

Tesco is a major UK-based retailer. Until recently, Tesco was a UK-only company, but it has expanded rapidly to become the world's third largest retailer behind Walmart and Carrefour:

● It operates in 14 countries (see Figure 6) and employs 500,000 people.
● It has 6300 stores worldwide with £65 billion in sales in 2012.

Tesco is a global retail brand. Most of its shops have the 'Tesco' name even though they differ in size and the range of products (Extra, Express, Metro, Superstore, One Stop). Abroad, Tesco stores often have a 'local' name, for example Tesco Lotus in Thailand and Tesco Kipa in Turkey. Tesco has concentrated on worldwide expansion in certain countries:

● Europe, because the market is close to the UK and consumers' tastes are similar
● Asia, because the market is so huge and incomes are growing, but there are few large supermarket competitors.

Tesco makes sure that it tailors its stores to suit local tastes, for instance, selling live fish in Thailand and dried jellyfish in China.

Like many retailers, Tesco has developed a large online business as a way of expanding by:

● delivering groceries via online ordering
● selling large items like furniture and electrical (Tesco Direct)
● introducing banking and insurance services (Tesco Bank).

Like many companies, Tesco has outsourced some of its administrative functions like data processing, stock ordering and accounting. About 6000 people work at the Tesco Hindustan Service Centre (HSC) in Bangalore, India, which opened in 2004. Tesco has done this because:

● wages for data processors in India are about 25% of those in the UK
● many Indians speak English, and there are many educated graduates
● HSC can communicate via the internet with the UK HQ, and all Tesco stores, very easily and cheaply
● Bangalore is known for this type of **outsourcing** so there are many trained workers available.

Figure 6 Tesco's worldwide expansion in Europe and Asia

Country	Opened	Stores in 2012
United Kingdom	1919	2,975
Hungary	1994	213
Poland	1995	412
Slovakia	1996	120
Czech Republic	1996	322
Republic of Ireland	1997	137
Thailand	1998	1,092
South Korea	1999	458
Malaysia	2002	45
Japan	2003	121
Turkey	2003	148
China	2004	124

Knowing the basics

Retailing is increasingly global, with similar brands and goods found in many countries.

Stretch and challenge

Jobs at Tesco are often low paid, but do provide many with part-time work and a 'first step' on the jobs ladder. Large retailers have become important employers in the UK.

Exam practice

Tested

15 What are the benefits of outsourcing to TNCs? [2]

16 Using examples of named TNCs, explain how they have expanded their global businesses. [8]

Answers online

Check your understanding

What is outsourcing?

Tested

Chapter 12 Development Dilemmas
How and why do countries develop in different ways?

Development means making progress so that people's lives improve. However, lives can improve in different ways – for instance, having more money, living longer, or knowing your children can get an education. Development is defined in different ways by different people:

Definition of development	Explanation
Economic development	An increase in the number of people working in the secondary and tertiary employment sectors – leading to rising incomes
Social development	Rising life expectancy, better healthcare and access to education; improved equality for women and minorities – leading to improved quality of life
Political development	Improving **political freedom** and the right to vote; a free press and freedom of speech – leading to greater control over who governs you

Development often focuses on **economics**. If people are getting richer, their lives must be improving – though this is not always the case:

- Living in polluted, congested cities might worsen health and increase stress.
- Incomes could rise, but so could the cost of living (food, water, housing), especially in cities.

Rather than think about development in a narrow way (e.g. income) it is better to think of broader **human development** – that is, improvements in income, health, education, equality, opportunity and freedom.

Globally, the human development situation has improved since 1970 as shown in Figure 1:

Figure 1 Trends in human development since 1970

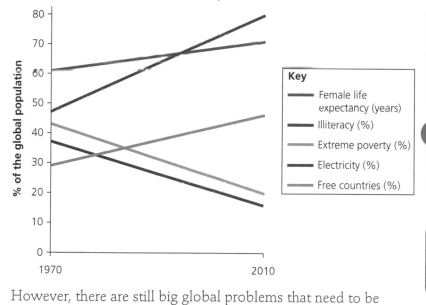

Key
- Female life expectancy (years)
- Illiteracy (%)
- Extreme poverty (%)
- Electricity (%)
- Free countries (%)

Stretch and challenge

The percentage of malnourished people globally is falling, but the actual number has been similar since 2000 because the global population is rising.

Check your understanding

What is meant by 'development'?

Tested

Knowing the basics

Development can mean economic, social or political improvements in people's lives.

However, there are still big global problems that need to be overcome. For instance, in 2012:

- 870 people were undernourished
- 1.3 billion people lived on less than US$1.25 a day
- 770 million people could not read or write
- 2.5 billion people lacked improved sanitation.

Exam practice

Tested

1 Describe the global development progress between 1970 and 2010 shown on Figure 1. [4]
2 What is social development? [2]

Answers online

Measuring development

Development can be measured in a number of different ways. The most common way is to use **Gross Domestic Product (GDP)** which is the value of all goods and services produced within a country in a year. Total GDP is divided by the population to calculate **GDP per capita** (per person). Three examples are shown below:

2011	UK	China	Uganda
GDP in US$	2,429,000,000,000	7,203,000,000,000	19,000,000,000
Population	63,181,775	1,354,040,000	34,131,400
GDP per capita US$	38,400	5,300	550
HDI (HDI range is from 0 to 1)	0.86	0.68	0.45

GDP per capita shows that the average Ugandan lives on US$550 per year, or US$1.50 a day. The average UK citizen has 70 times the annual income of an average Ugandan. GDP per capita is a good 'headline' way of comparing countries, but:

● it only considers income. It does not consider social development factors like health and education

● it does not recognise inequality within countries (there are 1 million millionaires in China, but 150 million people living on about US$500 per year)

● it does not consider the cost of living; goods are cheap in Uganda so US$550 is really worth about US$1300.

A widely used indicator of development is the **Human Development Index (HDI)**. It provides a broader measure of development than GDP per capita because it combines three pieces of data:

● life expectancy at birth (health)
● number of years in school (education)
● income per person (wealth).

Some countries are ranked higher by HDI than their GDP per capita. Cuba has an HDI of 0.78, but a GDP per capita of only US$6100: there is a very good health service and education system, but people are not wealthy.

Other ways of measuring development look at political freedom and **corruption**:

	UK	China	Uganda
2012 Corruption Perceptions Index (out of 100)*	74	39	29
2011 Democracy Index (out of 10)	8.2	3.1	5.1

*Transparency International's Corruption Perceptions Index measures the perceived level of public sector corruption in countries and territories around the world.

This is important because:

● in free countries (democracies) people can vote out a government that is not helping people develop
● a free press means that governments' bad decisions are exposed so people know the truth
● corruption means money that should be used for development is being stolen.

China scores poorly on the Democracy Index because there are no free elections and the internet is censored.

Check your understanding

Which is the broader measure of human development, GDP per capita or HDI?

Tested

Knowing the basics

It is better to use an index of development, rather than a single measure.

Stretch and challenge

China's low Democracy Index score does not mean China is not developing, but it does mean that ordinary people have very little say in how it is developing.

Exam practice

Tested

3 Compare the level of development in China, Uganda and the UK, using the data on this page. [4]

4 Explain why HDI is a more useful measure of development than GDP per capita. [2]

Answers online

The global development gap

The gap between the most developed countries and the least developed countries in the world is called the **development gap**. At its most simple the gap can be seen like this:

- In 2011, Luxembourg's GDP per capita was US$115,300.
- In 2011, Somalia's GDP per capita was US$112.

People in Luxembourg are on average more than 1000 times wealthier than Somalis.

The HDI 'gap' is just as large, ranging from Norway at 0.95 to the Democratic Republic of Congo at 0.29. Figure 2 shows that since 1970:

- developed countries, and some Asian **NICs**, have become much wealthier
- the gap between the developed and developing world has grown
- Africa has made very little progress in terms of rising incomes.

Figure 2 Changing income per capita

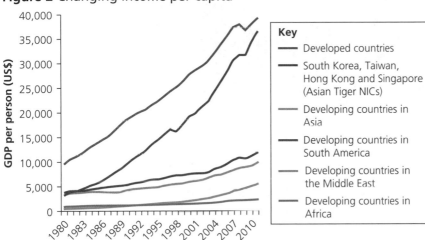

Key
- Developed countries
- South Korea, Taiwan, Hong Kong and Singapore (Asian Tiger NICs)
- Developing countries in Asia
- Developing countries in South America
- Developing countries in the Middle East
- Developing countries in Africa

Stretch and challenge

Inequality and poverty are not the same thing. People can be lifted out of poverty but the gap between them and the wealthiest can still grow.

exam tip

Learn some HDI and GDP per capita numbers for the exam as this will help you write precise answers.

There have been some very big success stories. In China average incomes grew from about US$800 in 1990 to over US$6000 by 2010.

It is also true that the world is very unequal:

- The richest 1% of people in the world have as much income as the poorest 50%.
- The 100 richest people have wealth greater than exceeds the annual GDP of the 40 poorest countries.

Inequality has not improved by a lot, as in 2007 the poorest 20 per cent of people still shared only 1 per cent of global wealth, just as they did in 1990:

% of wealth share by different income groups	1990	2007
Richest 20%	87%	83%
Middle 60%	12%	16%
Poorest 20%	1%	1%

Globally, the average HDI improved from 0.48 in 1970 to 0.68 in 2010. HDI has improved for many reasons:

- Aid has been given to developing counties, especially as part of the UN Millennium Development Goals since 2000, to help them improve health, education and water supplies.
- Many deadly diseases like polio, measles and yellow fever have been controlled in the developing world, causing life expectancy to rise.
- Countries in Asia (China, Malaysia, Thailand) have industrialised and this has increased incomes, so people have better diets and can afford healthcare.

Check your understanding

What is the global 'development gap'?

Exam practice

5 Using Figure 2, describe how incomes have changed since 1970. [4]

6 Using data from this page, describe the extent of the global development gap. [4]

Answers online

Knowing the basics

The world is very unequal and the gap between the richest and poorest remains very large.

Uganda: a Sub-Saharan African country

Revised

The developing world region that has made the least progress in terms of development since 1970 is Sub-Saharan Africa. Uganda is a country in this region. Like many developing countries, Uganda's economy relies on exporting of low value commodities – in 2010, 60 per cent of exports were food products, with coffee making up 20 per cent of all exports. Figure 3 shows that Uganda has made some progress, but war and changes in world coffee prices have knocked the economy off track in the past.

Figure 3 GDP per capita in Uganda since 1960

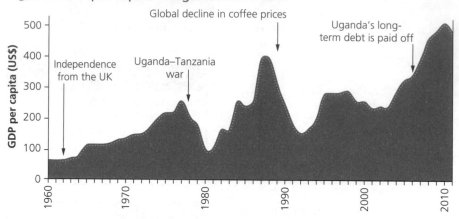

Knowing the basics

Many Sub-Saharan African countries have made slow and unsteady development progress.

Economic development	Human development	Political development
The economy has grown well since 2002. This is partly due to IMF reforms which have helped Uganda pay back debt and then use the money saved to invest in education and health	HDI rose from 0.31 in 1980 to 0.46 in 2012, but life expectancy only rose from 50 to 55 – partly because of the impact of HIV and AIDS in Uganda. Over 90% of primary school aged children attend school, but only 20% of those over 11 do so	Uganda was a brutal dictatorship in the 1970s under Idi Amin. Since the 1980s it has become more democratic, with political parties and elections, although human rights are still an issue

Further development progress in Uganda is likely to be affected by several barriers:

Landlocked	AIDS/HIV	Youth
Uganda has no coast and has to rely on other countries, like Kenya, for import and export routes	6.5% of adults in Uganda have HIV. This has fallen from 13% in 1990, it could rise if it is not managed	55% of Ugandans are under 18. They are a huge potential workforce, but also a social problem if they cannot get work
Corruption	**Conflict**	**Coffee prices**
Too much foreign aid still goes 'missing' in Uganda and TNC investors have to pay bribes to get business done	East Africa is plagued by conflict in nearby Kenya, Sudan and the DRC, which could spill over Uganda's borders	The economy is too dependent on a few crop exports; if coffee prices fall sharply, so does the GDP

Opportunities

Uganda has many strengths. Its youthful population could become a powerful driver of economic growth in the future. It has large areas of fertile farmland and can easily feed itself and export crops. Oil was discovered in 2002, and this could add US$2 billion a year to GDP, although agreement will be needed with Kenya to export the oil.

Check your understanding

Locate Uganda on a map of Africa and check you understand how it is 'landlocked'.

Tested

Exam practice

Tested

7 Suggest how Uganda's exports affect its economy. [2]

8 Explain the barriers to further development in Uganda. [6]

Answers online

How might the development gap be closed?

Theories of development

Two theories can help understand why societies develop over time. One is W.W. Rostow's modernisation theory (Figure 4) which dates from the 1960s. Rostow believed that development needed certain 'pre-conditions' before it could happen. These include:

- a move from farming to manufacturing industry
- trade with neighbouring countries
- the development of infrastructure like roads, ports and railways.

Once these happened a country would 'take-off' – it would develop secondary industry, begin to export and slowly increase trade and incomes.

Rostow's theory is based on what happened in Britain during the **industrial revolution**.

Rostow's theory has been very influential. Many countries have tried to get to stage 3 by investing in infrastructure, though they often borrow money to do so, which can lead to debt.

An important question is, why haven't all countries managed to move out of stage 1 or 2? The answer might be provided by dependency theory, an idea put forward by the economic historian A.G. Frank. Frank argued that **developed countries** exploit **developing countries** and this keeps them in state of 'underdevelopment' (Figure 5).

Figure 4 The five stages of Rostow's modernisation theory with country examples

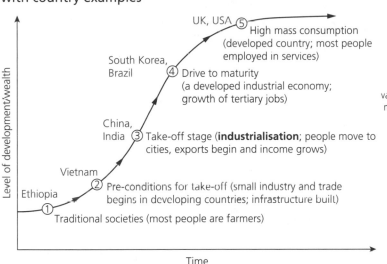

Figure 5 A.G. Frank's dependency theory

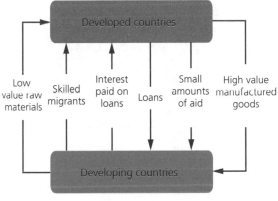

Developing countries provide developed countries with cheap raw materials (**cash crops**, metal ores), skilled workers and interest on loans they took out to try and develop. Developing countries depend on the developed world for costly manufactured goods, aid and loans (which lead to debt).

Dependency theory is rather simplistic. For instance, countries like China have broken free of this model and industrialised. However, some developing countries do seem to be too dependent on the developed world.

Check your understanding

Which theory does China's development fit best, modernisation or dependency?

Stretch and challenge

The two theories are linked; the least developed countries are often trapped in dependency but if they can begin to modernise and develop infrastructure they can break free and develop independently, like China.

Exam practice

9 Describe the stages of Rostow's modernisation theory. [4]

10 Outline two ways in which some developing countries are dependent on the developed world. [2]

Answers online

exam tip

You can draw models like Figures 4 and 5 in the exam, but keep any diagrams simple and clear.

Regional differences

Even within a country, levels of development vary so there are regional differences. A good example is India (Figure 6). In 2010 average per capita income in Bihar was £251 per year, but in Maharashtra it was £1,011.

Maharashtra had many initial advantages compared to Bihar which help explain the development difference. These include fertile soil and a good climate for farming, good water supply and a coastline which makes trade and transport easy.

Knowing the basics

Richer regions have often been richer for a long time, many centuries in some cases. Regional differences persist.

Figure 6 The location of Maharashtra and Bihar in India

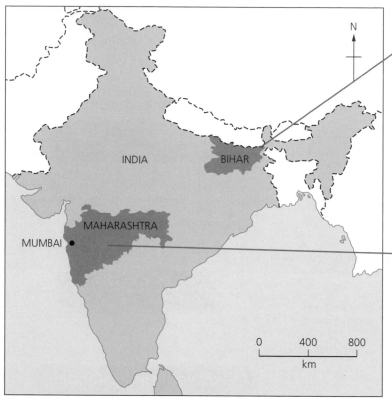

Eighty per cent of people in Bihar live in rural areas, making it part of India's **rural periphery**. Education is poor and the birth rate high. Poverty in Bihar is a result of:

● many people working as landless farm labourers

● very small farms that produce little more than one family's needs

● a very wide gap between rich and poor; Patna (the biggest city in Bihar) is one of India's richest cities

● government is more corrupt than elsewhere in India.

Maharashtra has three of India's largest cities, Mumbai, Pune and Nagpur, which are part of India's **urban core**. Mumbai in particular is a centre for:

● banking, insurance and call centres

● manufacturing industry from steel to textiles

● 'Bollywood', the centre of India's entertainment industry

● administration and government jobs which are often well-paid.

Stretch and challenge

There is often as much variation within regions as there is between them. Maharashtra has many very poor rural areas that are less fertile and more affected by out-migration of farmers than in Bihar.

Check your understanding

Compare how people earn a living in Maharashtra and Bihar.

Exam practice

11 Explain what is meant by an 'urban core'. [2]

12 Using an example, compare differences in development level within a country. [6]

Answers online

Top down or bottom up?

Revised

Development strategies are used by countries or regions that want to develop. Usually this means implementing a project that aims to improve aspects of people's lives. Development strategies can broadly be divided into **top-down** and **bottom-up**, which have very different characteristics:

	Top-down development	Bottom-up development
Aims	Economic development: these are large projects designed to improve incomes for many people, often by developing industry	Social development: these are smaller projects that aim to improve health education or food supply at a local level
Scale	Large: such as a whole region or a city	Small: a village, small rural area or an urban slum
Control	National: organised and controlled by central government in the capital city	Local: organised and controlled by the local community
Funding	Cost millions or billions of pounds and are sometimes financed by foreign loans and institutions such as the World Bank	Very low cost (hundreds or a few thousand pounds), often funded by outside **NGOs** such as Oxfam or Practical Action
Technology	Often highly technical, using imported machinery and foreign technical support	Often this is **intermediate technology**, which is simpler and needs less technical support, or renewable energy technologies
Examples	• Large hydroelectric power dams • Major roads, bridges and railways • New ports and airports • Major **commercial** farming developments	• Wells and water pumps • Schools and health clinics • Training for farmers • Village-scale energy projects such as biogas from cow dung

A bottom-up example is the UK NGO WaterAid which installs wells and hand pumps in Africa. These only cost £292 each, and use intermediate technology that can be maintained and repaired by local people. They provide clean water for a village, improve health and reduce the time women and children spend collecting water. However, thousands of wells are needed across Africa, so progress is slow for getting clean water to everyone. Clean water on its own does not improve people's incomes.

Check your understanding

Tested

Define the term 'bottom-up development'.

Knowing the basics

Top-down development is run by governments. Bottom-up development is run by local communities.

Stretch and challenge

Bottom-up development often meets basic needs, like health, water and food supply – but it may not increases people's incomes by very much.

Exam practice

Tested

13 Give one reason why bottom-up development strategies might be more suitable than top-down ones in poor rural areas. [2]

14 Outline the characteristics of top-down development strategies. [4]

Answers online

The impact of a large top-down project

Revised

China's Three Gorges Dam is the largest in the world. Like many top-down development strategies it was designed with the needs of the whole country in mind, to:

- produce hydroelectricity to help power China's industries and cities
- provide flood control on the lower Yangtze River below the dam in order to reduce flood damage
- improve trade with China's interior by allowing navigation on the upper Yangtze to Chongqing.

The Three Gorges Dam cost US$26 billion, took fourteen years to build and created a 600km-long reservoir behind the dam wall. It generates 22,500 MW of electricity.

However, building the Three Gorges has had many impacts:

- 1.3 million people had to relocate to make way for the reservoir.
- 1300 archaeological sites were flooded.
- Species like the Chinese river dolphin and Siberian crane are threatened by the dam.
- The reservoir may become polluted with farm and industrial waste.
- Rice farmers below the dam no longer benefit from annual flood waters irrigating their fields.
- The lake will silt up in 50 years and flood control will get more difficult.
- The dam is built on **fault** lines in an earthquake zone – catastrophic failure would put some of the 75 million people who live downstream at risk.

Every top-down project has winners and losers. This is true of the Three Gorges project:

Winners	Losers
People who get jobs at the dam and with the power companies, as well as cities upstream which benefit from increased trade	The 1.3 million local people who had to move from their homes, as well as any farmers on low incomes
People who run the Chinese and foreign companies managing the project, who made money from its construction	The fishing communities who find their fish stock either gone or reduced, both behind the dam and downstream
The people of China breathing in just a little less polluted air from the coal-fired power stations that would have been needed if the dam hadn't been built	The communities downstream protected from flooding have less available water for irrigation of rice fields, which has been fed by river flooding for centuries

Overall, cities and industry needing electricity in China seem to have benefited more than local people along the Yangtze River valley.

Check your understanding

Tested

What were the aims of the Three Gorges Dam project?

Knowing the basics

Remember that all development strategies have costs and benefits.

Stretch and challenge

Top-down projects often create problems for local people, but generate wealth and help economic development – the question is, which is more important? It depends on your perspective.

exam tip

Make sure you give a balanced view of impacts, and remember impacts can be positive and negative.

Exam practice

Tested

15 State two benefits of top-down development strategies compared to bottom-up strategies. [2]

16 Using a named example, examine the impacts of a top-down development project on different groups of people. [8]

Answers online

Section B Small-scale People and the Planet
Chapter 13 The Changing Economy of the UK
How and why is the economy changing?

Changing primary and secondary sectors — Revised

Over the last 50 years, the sectors of the UK economy have changed according to the **Clark–Fisher model** on page 79. The UK has moved from an industrial economy to a post-industrial one as the table opposite shows:

Employment in some UK industries has changed as the country has de-industrialised. There are many reasons for these changes, but two key ones are globalisaton and changes in government policy. Many UK industries were nationalised (taken into government ownership) in the 1940s and 1950s, but then privatised (sold off) in the 1980s as they had become loss-making.

UK labour force employment by sector	1960	1990	2010
Primary	4	2	2
Secondary	38	22	18
Tertiary	58	76	80

Industrial sector	Employment change 1960–2010	Explanation
Coal mining	Fell from 600,000 miners to only 6000	As UK mines became deeper, they became too costly to safely maintain Cheaper imported coal from Russia, Colombia and the USA The coal industry was government owned from 1946, but loss-making mines were closed in the 1980s and the industry was privatised in 1994
Iron and Steel	Fell from 250,000 iron and steel workers to 20,000	Suffered from strikes in the 1960s and 1970s and privatised in 1988 to prevent further losses to the government Steel made in South Korea and Europe was cheaper than steel produced in the UK
Footwear, textiles and clothing	Fell from 1 million in 1960 to 100,000 today	From the 1950s onwards, cheaper textiles were made in Taiwan, India and Bangladesh, often by using old mill machines exported from the UK **Globalisation** and cheaper transport costs made global exports and imports of textiles cheaper

It's not all bad news for UK secondary industry. In 1960 the UK made about 1.3 million cars. This fell to 900,000 in 1980, but in 2011 was back up at 1.3 million with about 75 per cent of all cars made being exported. The industry employs about 700,000 people. The big difference compared to 1960 is that most UK car factories are now foreign owned such as Nissan, Toyota, Jaguar Land Rover and Mini.

Check your understanding — Tested

What has happened to primary and secondary employment in the UK in the last 50 years?

Stretch and challenge

UK manufacturing today tends to specialise in high value, hi-tech products. The UK imports most manufactured **goods**.

Exam practice — Tested

1 Which economic sector do farmers and fishermen work in? [1]
2 Describe how employment in the primary and secondary sectors has changed in the UK since 1960. [4]

Answers online

Knowing the basics

Much of the UK's manufacturing industry could not compete with cheaper production overseas.

Changing tertiary and quaternary sectors

Most people in the UK work in the service sector. It's important to remember that this sector is very varied. It ranges from cleaners and shop workers on minimum wage (£6.19 per hour in 2012) to city bankers who can earn £250,000 or more in a year.

As the table below shows, tertiary jobs have grown. There have been big increases in health and education jobs. Most of these are in the public sector, that is they are services provided by the government. As the UK has become wealthier, more and more services are provided for people. In addition, retail, hotel and food employment has grown as wealthier people spend more in shops, eat out more and go on holiday more often:

Employment (millions)	Tertiary					Primary	Secondary
	Retail	Hotels & food	Finance	Education	Health	Mining	Manufacturing
1981	4	1.2	0.9	1.7	2.1	0.3	5.7
2011	4.8	2	1.1	2.7	4	0.06	2.5
% change 1981–2011	+20%	+66%	+20%	+58%	+90%	–80%	–56%

The table also shows that mining has almost disappeared as a job – in 2011 it only employed 61,000 people – and manufacturing employment has more than halved since 1981.

The number of workers in ICT, professional, technical and scientific employment grew from 1.7 million people in 1981 to 3.7 million in 2011. Many of these jobs are in the quaternary sector. This is an important sector because:

- the jobs are highly skilled, highly paid and employ university graduates
- research and development can invent new products that the UK can export
- many hi-tech companies are global **TNC**s that can quickly invest in new products
- many innovative companies are 'start-ups' set up by young entrepreneurs with a new idea or product that can be exported globally.

The UK has strengths in some quaternary industries such as:

- aerospace technology
- agricultural and food research
- chemicals, pharmaceuticals and biotechnology research
- digital imaging, internet and web design.

The **quaternary sector** is very competitive and other countries like China, the USA and South Korea are also part of the race to develop hi-tech industries and products.

Knowing the basics

Most UK people work in the tertiary sector, but the quaternary sector is also important.

Stretch and challenge

Quaternary employment is hard to get into – people employed in it usually have a university degree. It is high value but the number of employees is low, so it will never replace the secondary sector jobs the UK has lost.

Check your understanding

Tested

Which types of employment grew and declined the most between 1981 and 2011?

Exam practice

Tested

3 State two characteristics of jobs in the quaternary sector. [2]

4 Using the data in the table above, describe how employment changed between 1981 and 2011 in the UK. [4]

Answers online

Employment change

Revised

As well as big changes in the economic sectors people work in, there have been major changes in the nature of employment in the last 50 years:

	Change 1960–2010	Explanation
Total workforce	In 1960 about 24 million people were employed, growing to 27 million by 2010	The UK's population increased by 10 million, but more young people stay on in education until 18 or 21, and there are more elderly people who do not work
Average wages	Average weekly full-time wages in 2010 were £450, up from £14 (worth about £220 today) in 1960	Unskilled and skilled manual jobs are now rare. Most workers have some qualifications and skills; higher average earnings reflect this
Women in the workforce	Women make up 49% of the workforce now, up from 35% in 1960	Women receive over 50% of university degrees, up from 20% in 1960. Women are more independent, marry later on in life, have fewer children and focus more on careers
Part-time jobs	In 1960 only about 5% of people worked part-time, now 25% of people work part-time	The UK workforce is more flexible than in 1960, and people move jobs more often. Part-time work suits many people with families
Working hours	For all workers the average working week has fallen from 42 hours to 32 hours (37 for full-time workers)	The change partly reflects the fact that more people work part-time, and do some work at home; rigid '9 to 5' jobs are less common today

Back in the 1960s, many jobs were 'jobs for life'. This is not true today. Temporary work and part-time work are more common, and businesses increase or decrease staff numbers and hours to meet demand. During the recession that began in 2008, about 1.4 million people lost their jobs, as businesses cut back and often made full-time employees work part-time instead.

It is important to realise that people have different earnings depending on who they are, the sector they work in and their age:

2012 Average weekly earnings	Male full-time £506	Female full-time £449	Full-time aged 18–21 £280	Full-time manager £738
	Male part-time £146	Female part-time £158	Full-time aged 40–49 £572	Full-time retail £323

Check your understanding

Tested

How did the UK workforce change between 1960 and 2010?

Knowing the basics

There are more working women today, and average pay is higher than in 1960.

Stretch and challenge

While more women work today, they are still underrepresented in professional and managerial jobs – the so-called 'glass ceiling'.

exam tip

Discussing economic geography always benefits from some hard facts, such as wage comparisons.

Exam practice

Tested

5 State two reasons why there are more women in the UK workforce today, compared to 1960. [2]

6 Explain why the average earnings in the table above vary so much. [4]

Answers online

Contrasting regions

Revised

The UK is often said to have a 'North–South' divide in terms of its economy:

- The North was once heavily industrial, but these manufacturing industries have declined due to **de-industrialisation**.

- The South has an economy based on services (especially finance) and increasingly quaternary industries.

The North East (Tyneside and Teesside) and South East (Home Counties and south coast) regions have very different economic structures, as shown in Figure 1:

Figure 1 Comparison of industrial structure between the North East and South East of England

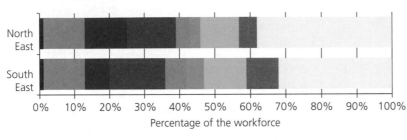

Key
- ■ Primary
- ■ Construction and transport
- ■ Secondary
- ■ Retail
- ■ Accommodation and food
- ■ IT
- ■ Finance and business
- ■ Scientific
- ☐ Public services

North East

South East

0% 10% 20% 30% 40% 50% 60% 70% 80% 90% 100%
Percentage of the workforce

Knowing the basics

The North East is still more dependent on secondary jobs compared with the South East.

Stretch and challenge

The North East is very dependent on public sector jobs, which have been cut back in recent years as the government looks to save money. There are few private sector jobs to replace the ones lost in the public sector.

The North East relies more on secondary industry employment and public services, whereas the South East has more private industry in areas such as finance, scientific research and IT. Economic conditions are much better in the South East:

2010 data	North East	South East
Unemployment	10.2%	6.1%
Average full-time weekly earnings	£443	£548
Working-age people with no qualifications	14%	8%
Population change 1971–2010	3% decline	25% increase

There are many reasons for these regional differences:

- The South East has become a centre for quaternary industry and research, especially along the M4 corridor which is located close to Heathrow, the national motorway network and London.

- Research and development spending in the South East in 2010 was £3.6 billion compared to only £0.3 billion in the North East.

- The South East workforce is better educated, so pay is higher and there are more jobs in science, technology and finance, whereas the North East has more low paid secondary and retail jobs.

Jobs in the North East are still being lost in traditional manufacturing, resulting in the high unemployment rate. The overall impact of the economic differences between the two regions can be seen in how their respective populations have changed. Overall, people have left the North East to move to the South East to look for work.

Check your understanding

How is the economy of the North East different to that of the South East?

Tested

Exam practice

7 State two reasons why the South East is a good location for quaternary industry. [2]

8 Compare the industrial structure of two contrasting regions you have studied. [6]

Answers online

What is the impact of changing work on people and places?

Environmental impacts of changing employment — Revised

As regions and urban areas move through the Clark–Fisher model (see page 79), the type of industry in these areas changes. The factories used for secondary industry close and are replaced by the offices, retail units and science parks of tertiary and quaternary industries. These changes have environmental impacts.

Sheffield was once the centre of the UK steel industry. This industry declined dramatically in the 1980s as a result of de-industrialisation with 120,000 manufacturing jobs lost from 1971 to 2008, a decline of 74 per cent.

Sheffield's industry has undergone **economic diversification**. Jobs have shifted from manufacturing into retail, software development, property development and business services. Sheffield has become a centre for specialist, hi-tech manufacturing.

These dramatic economic changes have had positive and negative environmental impacts on the city:

Derelict land ☹ About 900 hectares of derelict land and vacant buildings was created as steel factories closed, much of it polluted with heavy metals and other industrial waste	**Water quality** ☺ The River Don, which runs through Sheffield, was once grossly polluted and biologically dead in places. Factories polluted the river, but since these have closed, water quality has improved and the Don has been restocked with fish
Greenfield sites ☹ With less work in inner city factories, there has been increased pressure to build homes and businesses on greenfield sites of the edge of Sheffield	**Air quality** ☺ As the steel factories closed, air quality in the city improved as factory chimneys stopped polluting the air; nitrogen dioxide levels are about 50% of their levels in the 1980s
Traffic ☹ Jobs in Sheffield are no longer concentrated in the Don valley and city centre, so commuter traffic has increased as people travel further to their jobs, adding to congestion	**Regeneration** ☺ Some derelict industrial sites have been completely regenerated, such as the Meadowhall shopping centre which was once Hadfield's steelworks

De-industrialisation brings many environmental benefits, as heavy industry is usually very polluting. On the other hand, many jobs are lost. New jobs – often service or quaternary sector jobs – want to locate out of old, congested city centres and on greenfield sites at the edge of city, although this creates **urban sprawl** and the gradual loss of the countryside.

Knowing the basics

The closure of factories has many positive environmental impacts, especially if they were old and polluting.

Check your understanding — Tested

What is meant by 'de-industrialisation'?

Exam practice — Tested

9 Explain what is meant by 'economic diversification'. [2]

10 Outline the environmental impacts of de-industrialisation on a named urban area. [4]

Answers online

Stretch and challenge

The UK is less polluted because of de-industrialisation, but you should think about whether it would be better to still have those jobs. Could the polluting factories not have been cleaned up?

Brownfield and greenfield sites

Revised

The question of where new **economic development** should take place is a difficult one. There are two basic options: build offices, factories and retail parks either on brownfield sites or greenfield sites.

- **Brownfield sites** have been used before, but abandoned; regenerating them is rather like recycling land. Most are located in inner city areas.
- **Greenfield sites** have never been built on before; most are located around the edges of cities and are either farmland or woodland.

There are costs and benefits of developing both types of site:

	Greenfield	Brownfield
Size & shape	☺ Often large and regularly shaped; easy to build large buildings	☹ Often small and irregular, with existing land-uses on all sides
Construction	☹ All infrastructure (electricity, sewers, access roads) has to be built from scratch ☺ Construction costs are lower than brownfield	☺ Infrastructure already exists ☹ Might require decontamination of **pollution** ☹ Existing buildings need to be demolished adding to costs
Access	☺ On the edge of cities so there is good road / motorway access	☹ Often in inner cities with poor access on congested roads
Environment	☹ May destroy habitats if hedgerows, trees and ponds have to be removed ☹ Local people and environmentalists often object ☹ May encourage car travel as people commute to and from work	☺ As a form of recycling, it is more sustainable than greenfield ☺ Reduces the need for car travel as sites are in inner cities ☺ Improves the look of run-down areas and could increase property values

The UK headquarters of Sage Plc, the accountancy software company, opened on a greenfield site on the northern edge of Newcastle in 2004. Set in parkland, it is part of major city-edge housing and business development; the land was one of the largest parcels of UK **greenbelt** land ever to be released for development. Sage HQ is close to the A1 trunk road and only ten minutes from Newcastle International airport, an ideal location for a global TNC. Sage Plc sponsors The Sage Gateshead, a music venue built on former industrial land on the banks of the river Tyne in the CBD of Newcastle-Gateshead. This brownfield development attracted grants and lottery funding to help regenerate a derelict area. As a leisure and **tourism** attraction it makes sense to locate in the city centre.

Check your understanding

Tested

Where are greenfield and brownfield sites most often located?

Knowing the basics

New housing and industrial developments would usually prefer to locate on greenfield sites if they could.

Stretch and challenge

Local councils set targets for using brownfield sites, and restrict the number of greenfield sites available. This forces developers to regenerate brownfield sites.

exam tip

Make sure you have a named example of a brownfield site and greenfield site and how they have been developed.

Exam practice

Tested

11 What is meant by the term 'brownfield site'? [2]

12 Examine the advantages and disadvantages of brownfield and greenfield sites. [6]

Answers online

If the UK's post-industrial economy is to grow, it needs to take advantage of new economic opportunities. There are many of these, but the government needs to help them and private businesses need to invest in them. Some examples are:

The digital economy	East London Tech City (sometimes called 'silicon roundabout') is a cluster of 200 IT and technology companies including Google, Amazon and Intel. It is supported with £50 million from UK government, and aims to develop digital start-up companies for software, apps and games The government is investing £700 million to help 90% of people get access to superfast broadband, which should help the digital economy
Education and research	East London Tech City has partnered with Loughborough, Imperial and UCL universities to forge links between academic research and **commercial** development of digital products By encouraging young people to study engineering, science and computing at university, the skilled workforce the quaternary sector needs will develop The government is investing £2 billion in aerospace research from 2013–2020 and creating the UK Aerospace Technology Institute to promote aircraft and engine design by partnering with companies like Rolls-Royce
Green technology	Green energy, especially wind turbines and electric vehicles, is a key growth area. The UK had 8 gigawatts of wind power in 2012 and about 15,000 related jobs; wind power is expected to grow to 28 gigawatts by 2020 Nissan began producing the all-electric Leaf car in Sunderland in 2013, and this could be a future employment growth area
Green consumption	Interest in locally grown, organic food has increased in the UK. There are now around 800 farmers' markets where farmers sell direct to the public; this creates employment because many farmers now manufacture specialist products like ice cream, cheese and pies as well as grow the raw materials

The **green economy** could be a major area of growth for the UK, as people become more concerned about pollution issues and **non-renewable resources** eventually run out. The waste management and recycling industries in the UK employed 142,000 people in 2010. A lot of green technology is based on hi-tech research into new battery technology, energy efficiency and new lightweight materials. This is often developed with government money being used to set up partnerships between universities and private companies.

It is worth remembering that foreign workers can also be a source of economic growth, in three ways:

● Immigration can help plug skills gaps, when suitably qualified UK citizens are not available.

● Entrepreneurs with new ideas can come to the UK to set up new businesses.

● Lower-skilled workers can provide a low cost workforce and offset the UK's ageing population.

Check your understanding
Tested

What are the 'green' and 'digital' economies?

Knowing the basics

Many of the employment growth sectors require highly skilled and educated workers.

Exam practice
Tested

13 State two growth sectors of the UK economy. [2]

14 Describe how foreign workers could help the growth of the UK economy. [4]

Stretch and challenge

Remember that the UK has to compete with the USA, Germany and China in these growth sectors so it is not certain that many new jobs will be created.

Answers online

Changing working practices

As well as changes in the type of job people do in the UK, there have been changes in the way people work, who they work for and where they work. Perhaps the biggest cause of these changes has been the communication revolution.

These changes mean that people can work together, without the need to be in the same office, or even country, as the people they work with.

Type of work	Change
Teleworking (or telecommuting) means people work from home, or work 'on the road'	About 3.7 million UK workers sometimes work from home, with about 1 million mostly working from home (but who are not self-employed). In the 1980s this number was about 100,000
Home working means people are based at home all of the time	Home working has risen from 3.1 million people in 2001 to 3.8 million in 2011, with 66% of home workers being men
Self-employment means working for yourself (setting up your own business)	Self-employed people numbered 4.5 million in 2011, up from 3.3 million in 2002
Flexible working includes working part-time, job sharing, starting and finishing at different times	The number of part-time UK workers increased from 6.5 to 7.9 million during 1997–2011; about 40% work flexibly at some point in their career

These changes are likely to continue because:

● companies save money when people use their own home as a workplace, as large offices are not needed

● technology developments mean telecommuting and home working are likely to get even easier

● companies prefer flexible workforces, so they can easily shrink or grow their number of employees.

More flexible working does have positive and negative impacts.

Positive:

● Telecommuting reduces **commuting** and therefore saves fuel and creates less pollution.

● Part-time and/or flexible work does suit some people, like young parents and older people.

● People can choose when and where to work, fitting their work in with other commitments.

Negative:

● Being self-employed or 'freelance' brings with it the extra stress of looking for work all the time.

● There could be family tensions if homes are also workplaces for parents.

● There are fewer benefits like statutory sick pay or paid holidays.

● Part-time work pays less, so some people can struggle to get enough income to pay for living costs and their children.

Check your understanding

Do people work more or less flexibly than they did in the past?

Tested

Knowing the basics

Self-employment, working from home and working part-time have all increased in recent years.

Stretch and challenge

Many people would argue that employers are simply passing on employment risks and costs to flexible workers – which makes good economic sense but does not benefit the workers.

exam tip

Remember that impacts can be positive and negative.

Exam practice

Tested

15 Describe how working practices have changed in the UK in recent years. [4]

16 Using examples, explain the positive and negative impacts of more flexible working such as teleworking, self-employment and part-time work. [8]

Answers online

Chapter 14 Changing Settlements in the UK
How and why are settlements changing?

About 90 per cent of the UK's population lives in urban areas (**conurbations**, cities and towns). Liverpool and London are two urban areas that have had different fortunes over the last 50 years, as the data below shows:

Total population	1961	1971	1981	1991	2001	2011	Change 1961–2011
Liverpool	683,000	607,000	503,000	480,000	439,000	466,400	–32%
Greater London	7,800,000	7,500,000	6,600,000	6,900,000	7,200,000	8,200,000	+5%
Key ● Growing population ● Declining population							

Liverpool's population declined sharply between 1961 and 2001, but has slightly increased recently. London's population declined a little before 1981, but has since increased by 1.6 million. Why do these two British cities have such different patterns of population change? There are a number of factors to consider:

Factor	Liverpool	London
Economic	The city was very dependent on the dock trade and manufacturing industry, both of which declined sharply in the 1970s and 1980s	London's docks also closed in the 1970s and 1980s, but these were only a small part of the city's economy which included banking, tourism, business services and government jobs
Social	In 2010, it was found that 38% of working-age people have no formal qualifications – 10% higher than the national average	While London has many problems, it is perceived as a multicultural, global city with an educated workforce – attracting investors
Demographic	Many young people left Liverpool to find work, leaving behind an ageing, dependent population. Youth unemployment was over 60% in 1991	London has benefited from immigration of young people from around the world. In 2011, 37% of Londoners were born abroad and the average age was 37
Political	Many would argue Liverpool was ignored, especially in the 1980s when it was almost left in a state of managed decline	As the capital city, London has benefited from huge infrastructure investments like the M25, HS1 (a high-speed railway line to Europe), the London Docklands regeneration and the 2012 Olympics

Liverpool was shaken by globalisation whereas London benefited from it: Liverpool's port declined as manufacturing moved abroad, and small ships were replaced with large container ships too big for the city's docks. London's financial centre benefited from global growth in stock markets, it is connected to the world by Heathrow and Gatwick, and many global TNCs like HSBC and BP have their headquarters in London.

Knowing the basics

Population has declined in many northern industrial cities because of de-industrialisation, but London's population has grown.

Check your understanding

What has happened to the populations of London and Liverpool since 1961?

Tested

Exam practice

Tested

1 Using the data above, describe how London's population changed between 1961 and 2011. [2]

2 Using examples, explain why some UK urban areas have grown but others have declined in recent years. [8]

Answers online

Variations in quality of life

Revised

Even today there are stark contrasts between quality of life in UK cities. The most common way to measure this is using the **Index of multiple deprivation (IMD)**. This measures seven aspects of quality of life (income, employment, health, education, housing and services, crime, and environment):

- In 2010, 51% of all census areas in Liverpool were in the bottom 10% for multiple deprivation in the country – making it the most deprived city in England.
- In 2010, only 8% of census areas in London were in the bottom 10% for multiple deprivation.

Comparing Croydon (a London borough) with Liverpool reveals some differences:

2011 Census data	Liverpool	Croydon
Unemployment rate	7%	5%
Finance, IT and scientific employment	11%	20%
Born in the UK	90%	70%
Terraced houses/detached houses (%)	41%/7%	26%/12%
Council houses	27%	9%
No qualifications	28%	18%
Aged 0–15% / aged 60+%	22/17	17/19
People in bad and very bad health	9%	5%
Male life expectancy	74	79

Knowing the basics

Deprivation means lacking something that is considered 'normal' by most of society – like a decent income, a good education, a warm home and feeling safe.

Check your understanding

What is meant by the term 'deprivation'?

Tested

Knowing the basics

There are big variations in deprivation, both between cities and within them.

Compared to Liverpool, the population in Croydon is younger, healthier, better qualified, more likely to be in work and more likely to work in a higher paid job. Average weekly earnings in Croydon were £560 compared to £420 in Liverpool. In Liverpool people depend more on council housing and live in smaller terraced houses. People in Croydon have better services and opportunities:

- Croydon is connected to central London by the underground, rail and bus networks. People have the opportunity to travel quickly to find work across a wide area.
- Croydon was ranked 34th out of 152 council areas on its 2012 GCSE results, but Liverpool was ranked 104th. Young people in Croydon are better placed to find work than those in Liverpool.

Even the physical environment of the two areas is very different:

- Only 1% of land in Croydon is vacant or derelict compared to 7% in Liverpool (a result of factory closures).
- In Croydon, 37% of land is green space such as parks, compared to only 23% in Liverpool, meaning that people in Croydon have more recreational space despite living very close to central London.

There is some deprivation in Croydon, but de-industrialisation and population decline means there is much more in Liverpool.

Stretch and challenge

Living in Croydon is not all good; average house prices were £260,000 in 2013, compared to £125,000 in Liverpool, so getting on the housing ladder in Croydon is difficult.

exam tip

Make sure you understand that multiple deprivation is a composite measure – it uses seven different indicators of deprivation to calculate an overall score.

Exam practice

Tested

3 Using the data above, describe how Croydon is less deprived than Liverpool. [4]

4 Outline how access to recreational space could affect quality of life. [2]

Answers online

Changing rural settlements

Only 10 per cent of the UK population lives in rural areas. These are countryside areas with:

- low population density
- small settlements of under 10,000 people (small market towns, villages, hamlets and isolated dwellings)
- most of the land consisting of farmland, plus grassland and woodland.

Rural areas in the UK have changed dramatically in the last 50 years:

- Many areas are much less isolated than they once were because of new roads and a rise in car ownership.
- Large areas are protected and conserved as **national parks** and Areas of Outstanding Natural Beauty.
- Urban people visit rural areas more often because people have more leisure time.
- Farming is less important as a source of income. Tourism and leisure are now more important.

However, rural settlements are not all the same. Three distinct types can be identified:

Remote rural communities	Retirement communities	Commuter villages
Found in upland areas like the Lake District, the Pennines and Scottish Highlands. These locations are hard to get to, have poor transport links and often depend on farming	Attractive locations, often on the coast or on the edges of National Parks. Retired people 'downsize' their home and move here when they retire. Most locations are accessible areas close to towns	Found surrounding cities. These are **dormitory settlements** for people who commute to work by train/car but want to live in the countryside. They are very accessible
Often have an ageing, declining population	Have a stable, but aged, population	Have a rising population

- **Remote rural communities** have too few people to support many services. Many are protected areas that limit the amount of new development that can take place, and are too far from cities to be popular with tourists.
- **Retirement communities** in areas such as Devon, Cornwall, East Yorkshire and South Lakeland have developed because the UK population is ageing and people are living longer; retirees often have 20 or 30 years of retirement and want a peaceful and attractive location.
- **Commuter villages** are a result of **counter-urbanisation**, when middle-class families started moving out of the city to the countryside and commuting to work. Families like the safety and peace of the countryside, but still want to be close to the city and all its services. Commuter villages circle the UK's large cities, such as the Home Counties surrounding London.

Check your understanding

State three different types of rural settlement.

Knowing the basics

Not all rural areas are the same; commuter villages will have some 'urban' characteristics, like schools and shops, which are rarer in remote upland areas.

Exam practice

5 State two characteristics of a rural area. [2]
6 Explain why commuter villages have developed. [4]

Answers online

Stretch and challenge

Commuter settlements often suffer from a range of problems such as traffic congestion, high house prices and conflict between newcomers (commuters) and the original residents.

Contrasting rural regions

West Somerset and Hart District are two contrasting rural regions, at opposite ends of the deprivation spectrum:

- West Somerset, located in south-west England, north of Exmoor National Park, was the 45th most deprived local authority area in 2010.
- Hart in Hampshire (in south-west London) was the least deprived local authority area (326th).

West Somerset's population declined by 1 per cent between 2001 and 2011, whereas Hart's grew by 9 per cent. This contrast indicates that there is a demand to live in Hart, but not in West Somerset. Other data shows differences between the two locations.

Total population (2011 data)	Largest settlements	Average weekly earnings	Average house price	Superfast broadband availability
Hart District 91,700	Fleet (36,000) Yateley (20,000)	£841	£355,000	70%
West Somerset 30,075	Minehead (10,300) Watchet (4,400)	£523	£214,000	21%

A key reason for the differences in earnings and deprivation is employment. Five per cent of people in West Somerset work in farming and 15 per cent in accommodation and food, compared to 1 per cent and 4 per cent in Hart. These are low paid jobs. In West Somerset food and accommodation are linked to tourism which is very seasonal. Thirty per cent of jobs in Hart are in science, business and ICT, but only 15 per cent in West Somerset. These are high paid jobs. People in Hart are close to London, and the M4 corridor, so can easily work for large TNCs in high paid jobs.

Accessible Hart

- About 30,000 of Hart's residents commute out of the district each day, many to London, using the M3 motorway and commuter rail links.
- Car ownership and incomes are high which means people can access shops and other services with ease.
- It is easier for people in Hart to travel abroad as Heathrow, Gatwick and Southampton airports are all close by.

Isolated West Somerset

- Much of West Somerset consists of Exmoor National park, a beautiful but isolated upland area with only minor roads.
- There are no motorways in the area, and only three major roads. The average distance to a bank is 5 km.
- The area has no universities so many young people leave when they finish secondary school. West Somerset's population has an average age of 52, the highest of any local authority area in England.
- Transport and energy costs are high as there are few rural bus services and many people rely on burning oil central heating (21% of households were in fuel poverty in 2010).

Check your understanding

Tested

How do average earnings and house prices differ between Hart and West Somerset?

exam tip

Make sure you can name two contrasting rural areas and that you know some basic statistics for each area.

Exam practice Tested

7 Explain why deprivation levels are high in a rural area you have studied. [4]

8 Suggest reasons why isolated rural areas suffer from out-migration of young people. [2]

Answers online

How easy is it to manage the demand for high quality places to live?

Demand for residential areas — Revised

Cambridge is a historic university city about 60 miles north of London. In 2011 it had a population of 124,000. It is an area with very high housing demand because the population is growing. It increased 14 per cent between 2001 and 2011. Cambridge is also:

- a desirable, attractive place for families
- a good location for commuting to London by rail and the M11
- an area of economic growth, especially Cambridge Science Park which is a centre of quaternary industry
- a good location for university students, the population of which is growing.

The local council estimates that 14,000 new houses will be needed between 2011 and 2031. Housing supply in the city is restricted because:

- much of the city is a historic **conservation** area with listed buildings, so cannot be redeveloped
- the city is ringed by protected greenbelt land which cannot be built on.
- about 50% of all land in the city is protected green space, much of it owned by the university.

The shortage of land for new homes, and rising demand, means that house prices are very high:

Cambridge average house prices 2013	
Detached	£612,991
Semi-detached	£381,194
Terrace	£352,966
Flat	£276,667

High house prices have negative economic and social impacts. Young people cannot afford to buy their first home. It may also mean lots of people share a cramped rented house to reduce their individual rent. If housing demand is not met, businesses could decide Cambridge is not an ideal location because their workforce cannot afford to live in the city. High house prices are a problem for **key workers**. These are people like nurses, carers and council workers who are needed by society but who do not earn high incomes.

Cambridge has limited options in terms of meeting its housing demand, and most options have negative environmental impacts.

1. Greenbelt land: this could destroy habitats such as woodlands and hedgerows; it could also meet with objections from local people and environmentalists	**3. Urban green space**: for example, Midsummer Common, but locals would object if public space was destroyed, and the land is on the River Cam **floodplain**
2. Brownfield sites: Cambridge Regional College sold its Brunswick Road site and this has been developed as 'Cambridge Riverside' houses and apartments – but such sites are rare in Cambridge and will not be enough to meet demand	**4. Beyond the greenbelt**: for example, since 1998 4250 houses have been built at Cambourne, east of Cambridge, and up to 12,500 could be built on a former RAF site in Waterbeach to the north. The problem with this is that people commute to Cambridge and create traffic congestion and air pollution

Check your understanding

Tested ☐

Why is demand for housing in Cambridge growing?

Knowing the basics

Most urban areas in the UK do not have enough housing for everyone because the supply of land for new housing is too low.

Stretch and challenge

If key workers cannot afford housing near where they work, they are forced to commute long distances to work. This costs a lot of money and is a problem for people on relatively low incomes.

exam tip

You need to be able to name places and developments within your chosen urban area, not just name the urban area.

Exam practice

Tested ☐

9 State what is meant by the term 'greenbelt'. [2]

10 Outline the environmental impacts of rising demand for housing in a named urban area. [4]

Answers online

Rebranding and regeneration

Revised ☐

Cities are very dynamic places that are always changing. Sometimes, as in Liverpool, the change is negative, with declining industry and falling population. Cities need strategies to turn them around and attract businesses and people:

● Regeneration: this means physical redevelopment of an area with new buildings and infrastructure, so derelict and abandoned areas are brought back into use.

● Rebranding: this means changing the image of an area so outsiders see it in a more positive way, which helps attract investment and people.

Usually both strategies work together, as can be seen in Liverpool.

Liverpool's regeneration	Liverpool's rebranding
In 1981 the Merseyside Development Corporation began regenerating 320 hectares of the derelict Albert Docks into a maritime museum, Tate Liverpool art gallery, shops, and apartments. These opened in 1988	In 2002 Liverpool's airport (called Speke Airport) was renamed 'Liverpool John Lennon Airport' linking it to one of The Beatles, originally from Liverpool
Princes Dock has been regenerated and contains Liverpool Cruise Terminal (2007), the Malmaison and Crowne Plaza hotels and offices – all part of a £5.5 billion waterfront regeneration called Liverpool Waters	In 2004 the city waterfront became a UNSECO World Heritage Site, officially recognising the importance of the city's maritime heritage
Liverpool **central business district (CBD)** was regenerated with Liverpool ONE, a £900 million retail regeneration that opened in 2008, and includes shops, hotels, offices and a new bus station	In 2008 Liverpool became the European Capital of Culture; this prestigious award led to about £4 billion in investment and attracted 10 million extra visitors spending £750 million
Liverpool Knowledge Quarter is a hi-tech cluster centred on the Science Park (2006), Life Science Centre (2013) and Liverpool John Moores University	

Has Liverpool's rebranding and regeneration been a success? There have been some positive outcomes:

- Merseyside Development Corporation's regeneration created 22,000 jobs, built 97 km of new roads and footpaths, reclaimed 380 hectares of derelict land and attracted £700 million of private investment between 1981 and 1998.
- A further 25,000 jobs were created between 1998 and 2008, many in the culture, tourism and scientific sectors.
- Eighty-five per cent of Liverpool's residents felt the city was better after the Capital of Culture investment and events.
- Liverpool's population grew between 2001 and 2011 (it had been declining since the 1960s).

The improved image and physical infrastructure of Liverpool have halted the city's decline, but problems remain. It is still England's most deprived large city. Unemployment is higher than the national average and about one in three households in Liverpool have one unemployed member.

Check your understanding

Tested ☐

State two economic benefits of Liverpool's regeneration and rebranding.

Knowing the basics

Regeneration means improving the physical environment and economy of an area; rebranding means improving the image of an area.

Exam practice

Tested ☐

11 Using a named example, comment on the success of attempts to regenerate a named urban area in the UK. [8]

Answers online

Stretch and challenge

Liverpool's regeneration has helped the city avoid further economic and social decline, rather than transforming it into a city that can compete with London or other international cities.

Developing rural areas

In rural areas changes in employment can have dramatic impacts. The closure of a mine, or declining numbers of farm workers and fishermen can severely undermine the rural economy.

Many rural areas turn to tourism for economic development to replace the jobs lost in traditional industries. Tourism has problems:

● It usually depends on the **weather**.

● It is very seasonal, and so are the jobs.

● Pay is low (waitresses, shop assistants, guides) and skill levels are low.

● It is fickle – places that were popular last year may not be the next year.

However, there are other ways of sustainably developing rural areas, using larger projects and schemes:

Britain's Energy Coast, Cumbria

West Cumbria is a rural area in north-west England. It is an isolated, deprived area that includes towns like Maryport and Workington. These were once industrial centres which have declined. Plans for regeneration include:

● Britain's Energy Coast regional **development strategy** aims to develop the region by focusing on nuclear and renewable energy

● the creation of 3000 new jobs in the next 15 years

● Cumbria already has 18 onshore wind farms (106 turbines/96 megawatts) and five offshore wind farms (222 turbines/790 megawatts), with plans for more

● Sellafield, within the region, is the centre of the UK nuclear industry and new nuclear power stations could be built there.

● There is potential to develop the **biofuels** industry in the area.

The overall aim of Britain's Energy Coast is to develop the economy of West Cumbria and provide jobs for people in both rural and urban areas.

The Eden Project, Cornwall

The Eden Project is a major tourist attraction built in a former china clay quarry. This type of site is very difficult and expensive to reuse. It consists of **biomes** (for example, tropical rainforest, Mediterranean) housed in huge transparent domes. It opened in 2001 at a cost of £86 million. It has brought several benefits to the rural area:

● The Eden Project provides around 400 jobs and a further 200 seasonal jobs.

● Since it opened, over 13 million people have visited, bringing £1 billion into the local economy.

● Local hotels and pubs have benefited from the increased tourist trade.

● It is estimated that this has added 2500 jobs to the local economy.

● The ugly derelict quarry has been recycled into a unique visitor attraction.

● The Eden Project uses **rainwater harvesting** and has applied for planning permission to build a 4 megawatt geothermal power station (enough electricity for 4000–5000 homes), so it is a type of sustainable rural development.

The Eden Project is not without its problems. Traffic congestion was a major issue when it first opened. In 2013, 70 jobs were lost due to falling visitor numbers.

Both the Eden Project and Britain's Energy Coast could help stop out-migration. Both Cornwall and Cumbria suffer from young people leaving the region because there are few job opportunities.

Tested ☐

How can rural development projects stop out-migration from rural areas?

Knowing the basics

Large rural development projects are quite rare, but in areas of major rural deprivation they may be needed to create enough jobs to make a difference.

Stretch and challenge

Has the Eden Project been a success? Ten years after opening, the cracks may be starting to show as many people have already visited it and the recession means people are cutting back on spending – for most, it is a long way to Cornwall!

Exam practice

Tested ☐

12 State two economic reasons why many UK rural areas need rural development projects. [2]

13 Using the information on page 110, explain how the Eden Project can be considered sustainable. [4]

Answers online

Conserving rural areas

Revised ☐

When people visit rural areas they usually want to see a version of the countryside which is sometimes called the **'rural idyll'**. This means unspoilt, pretty, traditional villages with their pubs and village greens. Rural areas are often caught between two forces:

● Visitors and retired people want rural areas conserved to maintain their traditional character and landscape.

● Local people, especially the young, want economic development to ensure that there are jobs.

Planning policies have been used to conserve rural areas. Two examples are greenbelts and national parks:

	Greenbelts	National Parks
History	First allowed in 1947 as part of the Town and Country Planning Act	Set up in 1949 as part of the National Parks and Access to the Countryside Act
Aims	• To prevent urban sprawl • To preserve valuable farmland • To encourage reuse of brownfield sites in cities • To provide areas for recreation and leisure close to cities	• To conserve and enhance the natural beauty, wildlife and cultural heritage of rural areas • To promote opportunities for the understanding and enjoyment of the parks' special qualities by the public
Location	Lowland areas around major cities such as London, Manchester and Bristol	Mostly remote upland areas (Lake District, Snowdonia). Three main lowland areas (Broads, New Forest, South Downs)
Number	Fourteen main areas covering 1,639,560 hectares, or 13% of the land area in England	There are thirteen in England and Wales covering 11% of the land area. There are 110 million visitors to them each year

	Greenbelts	National Parks
Evaluation	• Planning rules are very strict in greenbelt areas so very little new development is allowed. This has been attacked as a very rigid policy that protects low value farmland rather than allowing land for new homes, which are needed as the UK's population grows • Greenbelts can be partly blamed for the UK's very high house prices • On the other hand, lack of greenfield sites close to cities means brownfield land is more likely to be redeveloped • Commuters often 'leapfrog' the greenbelt and develop new homes just beyond it – this means longer commuter journeys and some villages becoming dormitory settlements • Greenbelts have stopped urban sprawl and conserved rural areas	• National parks have strict planning guidelines for homeowners, farmers and businesses. This means very little new housing development occurs and new roads are rare • Because many national parks contain **honeypot sites**, pressures from tourism (litter, congestion, trespassing) are big problems • Many local people depend on seasonal tourism for a living • House prices are often very high due to second home buying, so locals cannot afford to buy • On the other hand, national parks have successfully conserved valuable rural landscapes

Check your understanding

Tested ☐

How are the aims of greenbelts and national parks similar?

Knowing the basics

Both national parks and greenbelts have been in place for over 50 years, and protect about 25 per cent of the countryside of England and Wales.

Stretch and challenge

Is protecting the countryside really more important than providing affordable homes for people? The national park and greenbelt policies, which date from the 1940s, might be seen as outdated.

Exam practice

Tested ☐

14 State two aims of the UK's policy of national parks. [2]

15 Examine the advantages and disadvantages of UK national parks for people and economic development. [8]

Answers online

Section C Large-scale People and the Planet
Chapter 15 The Challenges of an Urban World
How have cities grown and what challenges do they face?

Urbanisation means an increase in the proportion of people living in urban areas (towns and cities). In 2007, the number of people worldwide living in urban areas exceeded the number of people living in rural areas for the first time. We live in an urban world. However, the percentage of people living in urban areas does vary: from only 36% in Sub-Saharan Africa to 78% in the developed world.

Country/region	Percentage of people living in urban areas (2010 data)
Burundi	11%
Singapore	100%
Sub-Saharan Africa	36%
Developed world	78%
Developing world	46%

Trends in urbanisation are also different. Globally, the trend is upward, and 67 per cent of people are expected to live in urban areas by 2050. As Figure 1 shows, though, there are regional differences:

● In Europe and North America the number of urban people has more or less stabilised.

● In South America and Sub-Saharan Africa urban populations are still growing.

● In Asia, there has been enormous population growth in cities and this is expected to continue.

Figure 1 Urban populations by region 1950–2020

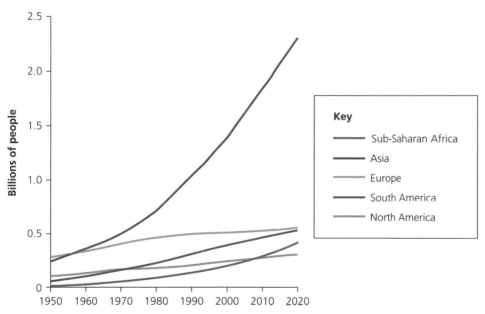

Key
— Sub-Saharan Africa
— Asia
— Europe
— South America
— North America

Cities grow for two mains reasons:

1 **Rural-urban migration**: people in rural areas migrate to cities to seek better opportunities, especially jobs, leading to urban population growth. Cities have a number of **pull factors** including higher incomes, more entertainment and perhaps better healthcare and education. Rural areas have **push factors** such as poverty, lack of farmland, tough physical labour and a lack of rural services.

2 **Internal growth**: population rises due to the high **birth rate** in the city.

Rural–urban migration and internal growth are linked. When young migrants arrive in a city they may start a family, and this pushes up the birth rate, adding to population growth. In developing world cities, about 60 per cent of urban growth results from internal growth and about 40 per cent from rural–urban migration.

In Europe and North America **counter-urbanisation** is taking place. This is where people migrate out of cities to live in rural areas – often **commuting** to work in the city. This is one reason why rates of urbanisation here have slowed.

Check your understanding
Tested ☐

State the two ways population in cities grows.

Knowing the basics

Urbanisation is happening most rapidly in the developing world.

Stretch and challenge

Cities are still growing quickly even in the developed world. Cities like Los Angeles and New York attract migrants as there are many job opportunities there.

exam tip

Make sure you can define the term 'urbanisation'.

Exam practice
Tested ☐

1 What is meant by rural–urban migration? [2]
2 Using Figure 1, compare the trends in urban population between different regions. [4]

Answers online

Megacities
Revised ☐

Increasing numbers of people live in **megacities** of over 10 million people. These are sprawling **conurbations** where cities have merged with surrounding towns, creating vast areas of urban landuse. The largest, Tokyo, houses 36 million people and covers 35,000 square kilometres – the same population as Canada and a similar area to Switzerland.

Globally,

● in 1975 there were only three cities with more than 10 million people (New York City, Tokyo and Mexico City)
● by 2010, there were 27 such cities and there are expected to be over 35 by 2030, with a total population of over 400 million people.

Most megacities are in the developing world and most megacity growth is occurring there:

	Developed world	Developing Asia	Africa	Latin America
Number of megacities	7	14	2	4
Average annual population growth	0.6%	2.8%	2.9%	1.4%

Megacities in the developed and developing world have very different economies. In the developing world, for example in Mexico City, many people have informal jobs, and the city has more manufacturing industry, construction and trade compared to New York City. As a global financial centre, New York City is dominated by highly-paid jobs in finance, business and other services such as law and the media.

Figure 2 Comparing megacity economies

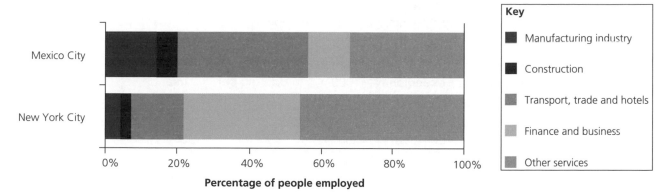

New York City's population grew from 19.7 million in 1990 to 22 million in 2010 (an increase of 12 per cent) whereas in Mexico City the increase was from 15 to 21 million (an increase of 40 per cent). The difference is because both rural–urban migration and internal growth are higher in Mexico City. In Mexico City 33 per cent of the population are under 20 years old, but this is only 26 per cent in New York City.

The two cities also contrast in terms of their spatial growth (the physical size of the urban area). New York City is only expanding slowly, as people continue to move into the suburbs. Mexico City is still expanding rapidly as more and more people move into the city and build around its edges. The city covered 1300 km² in 1990 but expanded to 2000 km² by 2010 (a 53 per cent increase).

Population density is usually higher in developing world cities. There are 9800 people per km² in Mexico City compared to only 1800 per km² in New York City (where many people live in large suburban houses).

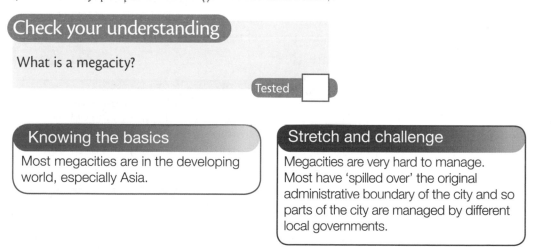

Check your understanding

What is a megacity?

Tested ☐

Knowing the basics

Most megacities are in the developing world, especially Asia.

Stretch and challenge

Megacities are very hard to manage. Most have 'spilled over' the original administrative boundary of the city and so parts of the city are managed by different local governments.

Exam practice

Tested ☐

3 Using Figure 2, compare employment in Mexico City and New York City. [3]

4 Suggest reasons why the population of some megacities is growing rapidly. [3]

Answers online

Urban challenges: the developed world

Most people who live in developed world cities have a high quality of life. There are areas of poverty and deprivation in these cities, but, on average, incomes are high and housing quality is good. This is because people have well-paid jobs.

To support this high quality of life, many resources are needed. These have to be brought into the city from surrounding areas. The table below compares average incomes, life expectancy and resource use in a developed world city and a developing world city:

	Income per person (US$)	Life expectancy	Cars per 1000 people	Water use (litres per person/day)	Electricity use (kWh per person/yr)	Waste (kg per person/yr)
Mumbai, India	1900	68	36	90	370	193
New York City, USA	56,000	78	209	600	6600	529

Quality of life is much higher in New York than Mumbai; people earn much more and live ten years longer on average, yet resource use is much higher in New York City.

Developed world cities are areas of concentrated resource **consumption** and this can lead to major challenges for the city in terms of supplying everyone with what they need. A good example is New York City:

- **Water**: water supply is 4.1 million m³ of drinking water per day, most from the north of the city in New York State and some from 200 km away.
- **Food**: this supply takes about 6 million hectares of farmland. Ninety-six per cent of all New York City's food is transported by lorry. A complex system of transport supplies 24,000 restaurants and 6000 food shops every day, but the city still wastes 200,000 tonnes of food each year.
- **Energy**: each year the city consumes 50,000 gigawatts of electricity, mostly from oil, gas and nuclear-fuelled power stations. The city's buildings use energy for lighting, heating, air conditioning and electrical devices.
- **Transport**: like most developed world cities, New York City suffers from severe traffic congestion. Many streets are not designed for cars and trucks and these vehicles' emissions mean that air quality is low, which increases health conditions such as asthma.
- **Waste**: the city produces 12,000 tonnes of household waste every day (only 17% of which is recycled) and another 13,000 tonnes from businesses: 90% of waste is transported by river barge to **landfill** sites.

Though the above sounds bad, New York City is quite a 'green' city by some standards. Few New Yorkers own a car because of the city's very high density, and efficient public bus and subway systems. House prices are very high so many people live in small apartments. Per person, **carbon dioxide (CO_2)** emissions are only 7.1 tonnes per year, which is far below cities like San Francisco at 11 tonnes.

Check your understanding

People in developed world cities consume large quantities of what types of resources?

Knowing the basics

Cities do not produce much of their own food, so this has to be imported (which uses up energy in transport).

Stretch and challenge

New York City may seem very wasteful in terms of resources, but as a compact city it is actually quite green. A New Yorker's eco-footprint is much lower than the average for the USA.

Exam practice

5 State two resources that cities in the developed world consume in large quantities. [2]

6 Outline some of the challenges of supplying resources to developed world cities. [4]

Answers online

In developing world cities, the challenges are often very different to those in the developed world. They mostly concern basic quality of life issues like housing, poverty and sanitation. Developing world cities often grow so quickly that the government simply can't keep up with demand for housing, water supply, power, transport and jobs. This leads to serious challenges:

Slums

- About 1 billion or 35% of people in developing world cities live in slums.
- Slums are low-quality housing without water and sanitation, often built by the inhabitants themselves.
- Most lack tenure, which means that residents don't own the land the slum is on, so they can be easily evicted.
- Slums are expected to house 2 billion people by 2030.

Informal economy

- In many developing cities most people work in the informal economic sector in jobs like recycling garbage, in small workshops or as domestic servants or street sellers.
- This means they pay no taxes and their work is unregulated.

The **informal economy** has a number of problems:

Income	Safety	Exploitation
Pay is low and work can stop at any time	Work is often dangerous and there is no insurance	Children often work, for very little money, in appalling conditions

In Mumbai 60 per cent of work is informal, but the 10,000 factories in Dharavi slum provide work for hundreds of thousands of people and generate over US$600 million in income each year.

Pollution

- Air **pollution** is high (Figure 3) because people cook on open fires, vehicles are often old and have dirty exhaust gases, and factory pollution is not regulated.
- Breathing in particulate matter (**PM10**) can cause asthma, bronchitis and even cancer – PM10 levels in Mumbai are dangerously high.
- Many rivers are polluted with sewage and industrial waste because cities cannot afford water treatment systems.

Figure 3 Air pollution in four cities

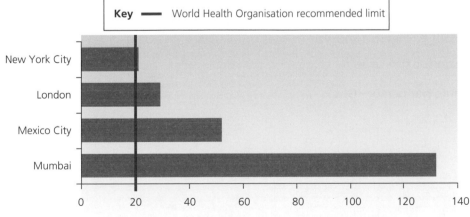

Key — World Health Organisation recommended limit

Annual average PM10 air pollution (micrograms per m³)

The combined effects of slum housing, pollution and the informal economy is low quality of life for many people in developing world cities.

Living in Mumbai

Mumbai in India has a population of 20.5 million people, growing at 4 per cent per year. It is a city of huge inequality with at least 20 billionaires but also 20 per cent of people living below the poverty line.

● About 54% of people in Mumbai live in 3000 slum areas across the city. The largest slum, Dharavi, houses 800,000 people.

● On average, people in Mumbai have only 4.5 m^2 of living space.

● The city has severe water shortages, with 650 million litres a day lost through old, leaking pipes. Some slum dwellers spend up to 20% of their money on water.

● Mumbai is very compact, so 55% of people walk to work. Only 2% of people own a car but it is still one of the most congested cities on earth.

● Three thousand people die crossing railway tracks or falling off packed commuter trains in the city each year.

Check your understanding

What are slums?

Tested

Knowing the basics

In developing world cities the key problems are to do with the low quality of life many people have.

Stretch and challenge

All cities are very unequal, but in developing cities the inequality is often very obvious; for instance, the contrast between squalid slums and ultra-rich gated communities.

Exam practice

Tested

7 Using Figure 3 on page 117, compare air quality in the four cities shown. [3]

8 Using a named example, explain why developing world cities often have low quality of life. [6]

Answers online

How far can these challenges be managed?

Urban eco-footprints

Revised

Ecological footprints (eco-footprints) measure the amount of land required to support people's lives in terms of their food, water and energy needs plus waste disposal. They are usually expressed in 'global hectares' per person (Gha). Eco-footprints vary from city to city, as Figure 4 shows:

Figure 4 Eco-footprints per person for five cities

	Eco-footprints per person (Gha)
Atlanta	13.0
Dubai City, UAE	11.8
London, UK	4.5
São Paulo, Brazil	4.4
Mumbia, India	0.7

The size of urban eco-footprints varies because:

- in developed cities like Atlanta, Dubai City and London, people are wealthier and consume more resources, especially energy, imported food, and through owning large houses
- car ownership is high in Atlanta and Dubai City. The cities are sprawling and low density so everyone drives, which uses more fuel; compact cities like London and Mumbai have lower eco-footprints
- in Dubai City, air conditioning is essential because of the very high temperatures, though this uses a lot of energy
- São Paulo is a very industrial city; despite it being a developing world city, its industry uses lots of resources which pushes its eco-footprint up to London's level
- cities like Mumbai are poorer and consume fewer resources, but are also better at recycling and produce less waste.

Many city governments, like London's, are aware that they are huge consumers of resources and that reducing their eco-footprints and carbon dioxide emissions is a priority.

London's annual eco-footprint

- 150,000 gigawatts of energy (equivalent to 13 million tonnes of oil)
- 50 million tonnes of materials and 7 million tonnes of food
- 900 billion litres of water
- 64 billion kilometres travelled, 70% by car
- 26 million tonnes of waste produced by households, construction and industry.

In 2013, Portland, USA had reduced the city's carbon dioxide emissions to below their level in 1990 by:

- using urban planning rules to prevent **urban sprawl**, leading to a compact, high density city, and pedestrianising the city centre – both of which reduce car (and energy) use
- the 2005 Green Investment Fund was a 5-year, US$2.5 million grant programme to support energy efficient building projects within the city
- support for the development of 'green' retailing – plastic bags are banned in large supermarkets, which reduces waste
- building a tramway system and providing cycle lanes and bike racks, both of which reduce car use
- the Clean Energy Works Portland programme provides grants for home insulation, which reduces energy use
- installing solar panels on 400 homes as well as several large **solar power** systems which reduce fossil fuel use.

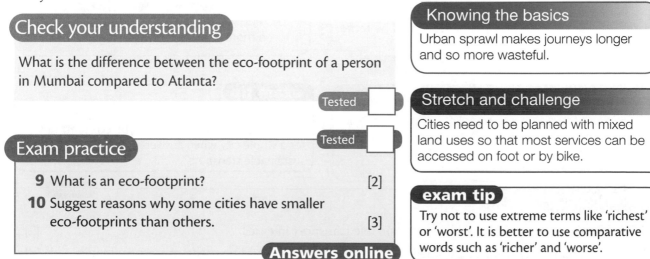

Check your understanding

What is the difference between the eco-footprint of a person in Mumbai compared to Atlanta?

Tested ☐

Exam practice

Tested ☐

9 What is an eco-footprint? [2]

10 Suggest reasons why some cities have smaller eco-footprints than others. [3]

Answers online

Knowing the basics

Urban sprawl makes journeys longer and so more wasteful.

Stretch and challenge

Cities need to be planned with mixed land uses so that most services can be accessed on foot or by bike.

exam tip

Try not to use extreme terms like 'richest' or 'worst'. It is better to use comparative words such as 'richer' and 'worse'.

Sustainable urban transport

One of the largest problems in developed world cities is transport. There are three main reasons for this:

- Cars, buses and trucks release harmful emissions which lead to poor air quality and can affect people's health.
- Petrol and diesel burn fossil fuels which increases a city's eco-footprint. Transport contributes 17 per cent to London's eco-footprint.
- Road traffic causes congestion which in turn causes delays and stress – 41 per cent of journeys made in London are by car.

London has attempted to make its transport more sustainable. This means making it:

- more environmentally friendly by reducing harmful emissions
- more efficient by reducing congestion, which reduces fuel use
- accessible to everyone by promoting low cost public transport.

London is aiming to reduce the city's carbon dioxide emissions by 60 per cent by 2025, partly by cleaning up transport. Examples of ways London has done this include:

Strategy	Aim	London's strategy	Comment
Congestion charging A tax on vehicles entering the city	To keep cars out of city centre by making it expensive, and so reduce congestion and pollution	In the Congestion Charge Zone vehicles pay £10 a day and there are fines of £60–£180 for people that don't pay. Electric vehicles do not need to pay	Traffic volumes have fallen by 16% since 2002 and pollution levels have reduced slightly, but there are parking problems just outside the zone
Encouraging bicycles Bicycles do not cause pollution and riding bicycles is good for maintaining health	To increase the use of bikes by building bike lanes and providing bicycles for people to rent in the city	In 2010 the 'Boris Bike' bicycle hire scheme was introduced. In 2012 there were 8000 bikes to hire from 570 'docking stations'. Four 'Cycle Superhighways', dedicated cycle lanes, have been built, with two more planned	The scheme was expensive to set up and does not make a profit. It costs a minimum of £2 to hire a bike and £90 for annual access – but it could become cheaper and profitable if it expands
Cleaner vehicles Electric and hybrid vehicles emit less pollution, and are quieter, which improves the environment	To encourage fewer polluting vehicles	London has introduced hybrid buses which emit 50% of the pollution of a diesel bus; London aims to have 1000 of these by 2016. London aims to have 1300 charging points for electric vehicles by 2013	This new technology is expensive as old buses have to be replaced, and people have to be encouraged to buy a new electric car (which can cost at least £25,000)

Other examples include low emission zones and integrated transport systems.

Check your understanding

State two ways that transport in cities could be more sustainable.

Tested ☐

Knowing the basics

Public transport is the key to making transport in cities more sustainable.

Stretch and challenge

It is hard to persuade some people to stop using their cars and to take a bus or a bike; this is why in London a tax (the congestion charge) has been introduced – but many people still choose to pay it.

exam tip

Make sure you don't confuse the average eco-footprint per person with the total eco-footprint for a whole city when answering a question about sustainable transport.

Exam practice

Tested ☐

11 State two problems caused by unsustainable transport in cities. [2]

12 Using a named example, examine how transport in a city could be made more sustainable. [8]

Answers online

Quality of life in developing world cities

Revised

The main challenges in developing world cities are social ones; for example, how to improve the quality of life for ordinary people. This can be done in a number of ways:

Self-help schemes in the Rocinha slum, Rio de Janeiro, Brazil

Strategy	Success?
Rocinha is a favela (slum) in Rio with a population of 100,000+. By self-help, the wooden shacks of the 1950s have been slowly upgraded to brick and concrete homes with water and electricity; people have rebuilt their homes a few bricks at a time when they could afford to. In some cases the city government and **NGOs** gave residents bricks and cement, and residents made their own improvements	Because Rocinha was never planned, it has no roads, only paths. This means access is poor and it is very cramped. The maze of paths is hard to police, and crime is a serious problem. Nevertheless, homes are much better than 30 or 40 years ago because of self-help, despite this being a very slow process

CORP (an NGO) in Mumbai

Strategy	Success?
The Indian NGO Community Outreach Programme (CORP) was set up in 1977. It runs 20 community centres in Mumbai, including in Dharavi slum. CORP's work focuses on education, helping street children, skills training for adults (jewellery making, tailoring), health and nutrition	NGOs can only help a small number of people and rely on donations for funding. In 2012 CORP spent £1.3 million and helped 29,000 people out of about 9 million slum dwellers in Mumbai

Urban planning in the Dharavi slum, Mumbai, India

Strategy	Success?
As far back as 1997 there have been plans to redevelop Dharavi, to demolish the slums and rebuild in a planned way. The most recent plan, the Dharavi Redevelopment Project, could cost US$3 billion	Dharavi's residents are against the redevelopment. They fear new apartments will be too expensive, and established businesses and factories will be forced to move. Residents who arrived in Dharavi after 2000 will not be rehoused

Urban planning in Curitiba, Brazil

Strategy	Success?
Curitiba is a city of 1.8 million people. Five rapid transit bus lanes radiate out from the centre, providing cheap single-fare bus transport. There are over 400 km² of parks and forest. The 'Green Exchange' programme gives free food and bus tickets to poor families in exchange for their garbage and recycling waste	Curitiba is a very 'green' city. Low income people benefit from the Green Exchange by using the cheap bus tickets to get to work, and by having a source of fresh food. It would be hard to apply this model to a vast megacity like Mumbai

Check your understanding

Who is mainly responsible for self-help schemes in slums?

Tested

Knowing the basics

Quality of life in developing cities can be improved though a combination of planning, NGOs and self-help.

Stretch and challenge

The scale of the slum challenge is huge; more people live in slums worldwide than live in the whole of Europe. Many developing world cities just do not have the money to tackle the slum problem.

Exam practice

Tested

13 State two features of a self-help scheme. [2]

14 Describe how NGOs could help improve quality of life in developing world cities. [3]

Answers online

Greener cities?

Revised

Mexico City was once one of the world's most polluted cities and was forced to take radical action.

It introduced the *'Hoy No Circula'* scheme in 1989. This means cars are banned from travelling on one day each week based on the last number of their number plate, for example cars with number plates ending in 5 or 6 cannot be used on a Monday. The scheme was extended to Saturdays in 2008. As a result, air pollution has fallen and is now similar to pollution in Los Angeles (whereas in 1992, the UN named Mexico City 'the most polluted city on the planet'). However:

- wealthy people have been accused of second car ownership, to get around the restrictions
- because the city is still growing rapidly, the number of vehicles is still rising
- Mexico City has recently invested in better public transport, especially its Metro system, which has had a bigger impact on reducing pollution and congestion.

Another approach to developing less polluted cities is to build new cities from scratch. Building new cities is not cheap and the plans are inevitably affected by economic and political change.

- Dongtan, the eco-city planned by the Chinese outside Shanghai, has made virtually no progress. Dongtan consists today of ten wind turbines – no buildings, water taxis, water cleansing plants or energy centres. Construction was to have started in 2006 but nothing has happened yet.
- Masdar, the eco-city planned in Abu Dhabi, has met with considerable delays and major changes to the plans, for example, abandoning self-sufficiency in solar energy. Critics have suggested that, if completed, it will become the 'ultimate' gated suburb occupied by a rich elite who carry on very unsustainable lifestyles outside the city.

These radical attempts at less polluted, greener cities have advantages and disadvantages:

Check your understanding

What does Mexico City's *'Hoy No Circula'* plan mean?

Tested

exam tip
Questions that ask you to write about how cities can become more sustainable are usually focused on what planners can do.

Advantages	Disadvantages
Greenfield development: planned from scratch so it is possible to design very efficient transport, water and waste systems. These are costly and difficult to install in existing cities **Test-bed**: new cities can be used to try out and adapt new low-pollution technology, which can then be used more widely **Quality of life**: air quality, housing quality, rapid transport and public services can all be designed into a new city to ensure everyone has a good quality of life	**Cost**: Masdar is expected to cost up to US$20 billion, but can only house 50,000 people **Complexity**: many new cities use untested technology which is expensive and does not always deliver what it promised **Land**: an undeveloped area of land is needed and people nearby object to the huge new development – this is one of the reasons the UK's eco-towns failed

Stretch and challenge
In most cases, new cities are simply too expensive so grand plans rarely get built.

Knowing the basics
City planners have many opportunities to reduce their city's eco-footprint, including managing transport, water, and waste, and keeping to building regulations.

Exam practice
Tested

15 State two strategies that cities can use to reduce urban pollution levels. [2]

16 Using named examples, explain the advantages and disadvantages of attempts to develop less polluted cities. [8]

Answers online

Chapter 16 The Challenges of a Rural World

What are the issues facing rural areas?

Rural areas in the developed world ———————— Revised

Rural areas are very different to urban areas. They have:

- low population densities and low building densities
- small settlements (farms, villages, market towns) rather than large towns and cities
- open space, fields, woodland and wild areas like mountains.

People also earn their living differently. In the UK's Peak District (a rural area in Derbyshire), farming, manufacturing and **tourism** (accommodation and food) are much more important to the rural economy than they are in the UK as a whole (Figure 1). Many rural areas in the UK increasingly depend on tourism, leisure and recreation.

Living in a rural area has some advantages, such as an attractive landscape, low crime rates and a more relaxed pace of life compared to urban areas. However, services are often much less accessible than in urban areas. It may be many miles to the nearest supermarket or hospital. Rural areas have

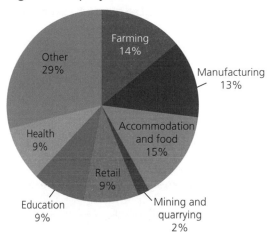

Figure 1 Employment in the Peak District

- Farming 14%
- Manufacturing 13%
- Accommodation and food 15%
- Mining and quarrying 2%
- Retail 9%
- Education 9%
- Health 9%
- Other 29%

low populations spread out over large areas and so cannot support **high order services**.

In rural areas in the developed world, farming is **commercial**. This means farmers aim to sell produce and make a profit. There are several different types of farming found in different locations:

Arable farming (growing crops)	Pastoral farming (rearing animals)
Lowland arable Wheat, barley and potatoes. Found in the drier and warmer east of England	**Hill sheep farming** Found in upland areas like the Lake District and Mid-Wales
Horticulture Vegetables and fruit. Found in the south and east, often close to major cities	**Dairy farming** Milk. Found in the wetter west of Britain, but in lowland areas

Arable and dairy farming are usually very **mechanised**. Because UK farming is commercial, there are large inputs to make sure **yields** are high. These include water for irrigation, farm chemicals (fertilisers and pesticides) and diesel for the machines.

exam tip

Make sure you know some examples of different types of farming when answering a question on this topic.

Stretch and challenge

Few UK farms make much money. Only very large, specialised farms are very profitable. Many farmers struggle to get by on low incomes.

Exam practice — Tested

1 What is meant by commercial farming? [2]
2 Describe how people earn a living in rural areas in the developed world. [4]

Answers online

Check your understanding

State two ways rural areas in the developed world are different to urban areas.

Tested

Rural areas in the developing world

In the developing world, the rural economy is even more dependent on farming in terms of incomes and people's survival. In Uganda, Africa, 86 per cent of people live in rural areas (compared to 20 per cent in the UK) and 82 per cent of people work in farming (compared to less than 2 per cent in the UK).

In rural Uganda, 85–90 per cent of households depend in farming as their main source of income. Most farming here is **subsistence farming**. This means:

● households grow food and rear animals to feed themselves

● Households may grow some crops to sell, and sell any surplus food, to earn money.

Farming in Uganda is very different from farming in the UK:

● Most farms are small family plots, 1–3 hectares in size (1 hectare is about the size of a football pitch).

● Only 1 per cent of farms in Uganda use chemical fertilisers or irrigation water.

● Labour is done by humans rather than machines.

● Most farms are mixed, with households both growing crops and keeping animals.

Crops	Animals
Ten crops dominate farming: matoke (a kind of banana), beans, cassava, sweet potatoes, coffee, groundnuts, maize, millet, sorghum, and sesame.	Twenty per cent of farm households own at least one cow; 30 per cent keep goats and 46 per cent keep chickens.

The only **cash crop** grown by small Ugandan farmers is coffee, but this accounts for 93 per cent of all coffee grown. Coffee is grown in small quantities and sold for money. The coffee is exported to countries like the UK. Maize is often sold locally if a farm manages to produce some surplus.

There are some large commercial farms in Uganda. These employ an average of 70 workers and grow crops for export, mainly tea and tobacco. These cash crops are very important to the economy of Uganda. About 60 per cent of all of Uganda's exports come from farming and, as Figure 2 shows, four of these are very significant:

Subsistence farming in Uganda is a difficult life. One-third of rural households live below the poverty line, and about 50 per cent of income is spent on food and drink. Despite most people growing their own food, they still need to buy some food.

Uganda's farmers get very little money even for their main cash crop, coffee. The coffee is quite low quality and they sell it at a low price because the cost of transporting it out of isolated rural areas is so high.

Figure 2 Percentage of Uganda's exports in 2010 from four cash crops

Coffee beans	29%
Tobacco	9%
Cocoa beans	5%
Tea	5%

Check your understanding

State two characteristics of farming in Uganda.

Tested

Knowing the basics

Many subsistence farmers live just above the poverty level; if their crops fail they can be plunged into poverty.

Exam practice

Tested

3 Explain what is meant by subsistence farming. [2]

4 State two examples of cash crops grown in the developing world that are exported to the developed world. [2]

Answers online

Stretch and challenge

Unusually, Uganda's cash crop, coffee, is grown by small farmers; in many developing countries cash crops are grown on large commercial plantations.

The Lake District **National Park** (LDNP) in Cumbria, north-west England, is a protected rural area with:

- an area of 2,292 km²
- a resident population of 40,770 people
- 14.8 million visitors a year.

It is a beautiful landscape, but a very rural one. There are three towns in the LDNP: Keswick, Windermere and Ambleside. Despite its beauty, the LDNP faces a number of challenges:

Issues	How is it a rural issue?	Details
Rural **isolation**	While towns are quite well served by services, villages and farms are not. Those living there need to either own a car (and spend money on fuel) or rely on public transport	For old people especially, the time and cost of travel is significant. The LDNP only has three main roads running through it and infrequent bus services, especially in winter (outside the tourist season) and at weekends
Changing employment	In rural areas, employment in farming has declined. This is because machines have replaced humans and cheap food can be imported from aboard. Tourism has increased because of increased mobility and leisure time	There are only 2500 people employed in farming in the LDNP, and net farm income is only about £8000 – not very attractive. Forty per cent of LDNP jobs are in hotels and restaurants and are poorly paid, seasonal and part-time
Housing and counter-urbanisation	Many rural areas face a shortage of affordable housing. This is because wages are low and there are few **brownfield sites** where development would be easy. Counter-urbanisation pushes up house prices as commuters buy houses in rural areas	In the south of the LDNP, close to the M6, house prices are very high and house shortages acute. As a protected area, planning permission for new homes is hard to get
Tourism and leisure	Leisure and tourism helps provide jobs and income. However, it also brings traffic congestion, and tourism changes the types of services and shops from local to tourism-related ones	Traffic and visitor congestion in the summer is a problem as 90% of visitors arrive by car. Tourism employs 15,000 people and is worth £900 million a year. Footpath **erosion** and litter are serious issues in tourist 'hotspots'
Changes to rural services	Rural services in villages have tended to close because many people travel by car to towns to use services. Services are expensive to run, so are the first to be cut back	Nearly 70% of people outside of the main towns live more than 3 km from a post office. There are only three high schools within the LDNP boundary and one hospital.

In 2008, LDNP residents were asked which issues mattered to them most. Sixty-four per cent of people said lack of affordable housing was an important concern, 42 per cent said lack of activities for teenagers, 40 per cent said lack of public transport and 38 per cent said low wages and the cost of living. Lack of rural services, affordable housing and low incomes were the main issues.

Stretch and challenge

Because national parks like the LDNP are **conservation** areas, farmers and local people are restricted in how far they can develop farms, buildings and new businesses, as there are strict planning rules.

Check your understanding

State two local issues that people in the Lake District are concerned about.

Tested

Exam practice

Tested

5   What is meant by rural isolation? [2]

6   Using a named example of a rural area in the developed world, describe how its economy has changed. [4]

Answers online

Knowing the basics

Rural areas in the UK have changed a lot in the last 50 years; farming is less important and tourism and leisure are more important.

Rural challenges in a developing country

Western Uganda is an area typical of developing countries in Africa. It is rural, very isolated and economically dependent on subsistence farming. A number of challenges affect the people who live there:

Issue	Explanation	Impact on people
Rural isolation	Only 9% of rural communities are connected to an electricity supply. Roads are very poor, with only 25% paved, meaning that 75% of people live two or more hours from a market. The average distance to a health clinic is 7 km and 75% of people walk to these clinics	Access to basic services is poor, which restricts rural development. Even something as simple as taking crops to sell at a market can take all day. Isolation prevents farming becoming more commercial and profitable because market access is so poor
Changing landholdings and farm economy	Most rural households do not have tenure of the land they farm; this means it is hard to prove they have a right to live on it	Evictions from land are common; in 2012 8000 people were evicted from land because of a new oil refinery being built close to Lake Albert. In 2011 about 20,000 people were evicted from land by the New Forest Company who wanted to plant forests
Rural–urban migration	Poverty and lack of services means that many people migrate from western Uganda to Kampala, and increasingly to the oil fields close to Lake Albert, to look for work. Most migrants are men	Kampala, the capital city, is growing at 5.6% per year and its slums are expanding. Women are left in rural areas with even more farm and domestic work to do, which risks increasing food insecurity due to labour shortages
Natural hazards	Uganda, like many developing countries, experiences natural hazards such drought and floods (see Figure 3). Drought is the most common natural hazard	Drought and flood directly destroy crops, which people depend on for food. People in developing rural areas are vulnerable to natural hazards because they grow their food, and have no other food or income source

Figure 3 Natural disasters in Uganda

Disaster	Date	Number of people affected
Drought	2008	1,100,000
Flood	2007	718,045
Drought	1999	700,000
Drought	2002	655,000
Drought	2005	600,000

Knowing the basics

Rural areas in the developing world are very isolated. It could take as long as a whole day to get to the nearest town.

Stretch and challenge

Land ownership, and being able to prove you own the land you farm, is a key issue in developing world rural areas. Most people do not have a piece of paper saying they own their land. This makes them vulnerable to eviction.

Check your understanding

Tested

State two problems faced by people in rural western Uganda.

Exam practice

Tested

7 State two reasons why some people in rural areas in the developing world choose to migrate to cities. [2]

8 Describe how people in rural areas in the developing world are at risk from natural hazards. [3]

Answers online

How might these issues be resolved?

The role of different groups in rural areas in the developing world

In developing countries there is often a focus on trying to help people in rural areas by improving their livelihoods and giving them opportunities to develop economically. Rural development projects often focus on:

- developing agriculture to improve food supply and increase incomes

- ensuring a safe and secure water supply
- developing infrastructure to improve trade and reduce isolation
- providing basic services like health clinics and education to improve quality of life.

Different groups have different roles in delivering rural development projects:

Inter-Governmental Organisations (IGOs)	Global organisations like the World Bank or United Nations. • IGOs often provide a framework of targets for countries to achieve, like the UN Millennium Development Goals (MDG) which run from 2000 to 2015. The MDG aim to reduce poverty and hunger by half by 2015. • IGOs often provide funding for rural development, but usually for large-scale projects like rural road networks or electricity grids.	**Top-down development**
National government	• National governments often implement nationwide networks such as primary schools or health clinics. • They tend to be focused on targets – such as getting all children into school or vaccinating a certain percentage of people.	
Local government	This is the government of a region or small area, like a council: • In many developing countries, local government is quite weak and poorly financed. • In some cases, such as Kerala State in India, the local government successfully focused on improving literacy, especially for females.	
Non-Governmental Organisations (NGOs)	They can be large global charities like Oxfam or Practical Action, or smaller local ones: • NGOs rely on donations. • They tend to work with local communities on small-scale projects to provide basic needs and improve incomes. • They can often only help small numbers of people, but do so in an intensive way.	**Bottom-up development**
Local communities	• NGOs and local government often encourage local communities to take control of development projects once they have been set up. • Community councils or women's groups might manage, maintain and adapt projects. Some rural development is bottom-up. This means the local community has a role in deciding how development projects should work. Other groups are more top-down in their approach, meaning rural development is imposed from above on rural communities.	

Check your understanding

Tested

Name an organisation that carries out top-down development and one that carries out bottom-up development.

Knowing the basics

Rural development projects can be carried out by global organisations like IGOs or very local NGOs.

Exam practice

Tested

9 What is an NGO? [2]

10 State one advantage and one disadvantage of rural development projects carried out by NGOs. [2]

Answers online

exam tip
Know some named examples of NGOs and IGOs.

Improving rural lives in the developing world

There are a wide range of rural development projects designed to improve quality of life and opportunities in rural areas of the developing world.

Biogas systems
These have changed rural life in India.

The problem
- Firewood fuel is in short supply, with 30 kg needed each week for each family for cooking.
- Cow dung can be used, but it would be better used as fertiliser on fields. Burning it causes health problems.
- Women and girls collect fuel. The time they spend doing this could be spent in school.

The solution
Each biogas 'digester' costs between US$500 and US$1000. Cow dung and human waste is used. Both cooking gas and gas for lighting can be provided. There are 4 million of these in rural India.

- Biogas digesters produce methane gas and 'slurry' for fertiliser.
- They save time, allowing girls to go to school, which helps them later on when getting a job.
- They are relatively easy to construct and cheap to maintain.
- They improve health because less smoke is inhaled.
- They create jobs – building and maintain biogas plants employs about 250,000 people in India.
- They can be used to generate electricity to pump water.
- They can provide gas lighting at night so people can work and study after dark.

Despite these benefits, they are still too expensive for many rural communities in India to afford.

Micro-finance
An example of **micro-finance** is Grameen Village Phone in Bangladesh:

- The NGO Grameen Bank provides small US$200 loans to women in villages in Bangladesh.
- They use the loan to buy a mobile phone, and are trained how to use it.
- Other villagers pay the women to use the phone; they back the loan and make a small profit.

The mobile phones allow villagers to trade by checking prices before they go to market, keep in touch with relatives who have moved to the city, and receive health advice over the phone. Micro-finance can also lead to debt, despite the loans being very small.

Mobile healthcare
Riders for Health is an NGO that provides mobile healthcare in Africa. It uses motorcycles with side-cars (that can carry a stretcher) because they can cope with Africa's poor roads and be maintained easily. Riders for Health:

- has 1400 motorcycles in six African countries
- has 50 motorcycles in Kenya which reach 20,000 people per week
- provides vaccinations, care for malaria and HIV, health education and maternal care.

Riders for Health reaches rural people that normal health services would miss because of isolation. It is low cost, and effective, but like many NGOs it is small and has to rely on donations. This means it can only reach a fraction of the people who need healthcare in rural Africa.

Knowing the basics
Rural development projects in the developing world usually focus on meeting people's basic needs like healthcare, clean water and food supply.

Check your understanding

State two benefits of biogas digesters in India.

Tested

exam tip
When discussing development projects always try and use named and located examples.

Stretch and challenge
All development projects have to be financed by someone, and for this reason millions of people never get this sort of help.

Exam practice
Tested

11 What is meant by micro-finance? [2]

12 Using named examples, examine how development projects can improve quality of life in rural areas in the developing world. [8]

Answers online

Farm diversification in the developed world

In the developed world, farmers are increasingly focusing on **diversification**. This means using their farms to develop new sources of income, other than growing crops and rearing animals. The reasons for this are:

- Falling farm employment and the rise of cheap imported food.
- In the past, farming was subsidised by the European Union, but these payments have fallen.

- Farming often does not bring in enough income to support a family.
- With more people visiting the countryside there are opportunities to make money from leisure, tourism and recreation.

About 50 per cent of England's farms have diversified in one way or another. Many have several types of diversification on one farm. The most popular ways are:

Type of diversification (2009)	Percentage of farms	Number of farms	Total income for all farms (£ millions)
Renting out farm buildings for other uses such as shops or as business units	36%	21,000	1400
Making specialist food on the farm (ice cream, pies) and farm shops	7%	4200	170
Incorporating sport and recreation such as paintballing or a campsite	11%	6500	330
Providing accommodation for tourists such as holiday cottages	5%	3100	90

Seal Stoke Farm in Devon diversified because wool from their sheep was fetching only £1 per fleece. The farm diversified into Texel sheep, whose fleece can be made into natural wool filled duvets. A local mill turns the fleece into filling and local seamstresses create natural cotton covers. Both of these provide local people with jobs. The farm also has a self-catering cottage for tourists.

About 20 per cent of farms earn more from diversification than just from farming, but there are issues:

- Isolated farms find it harder to make a profit than farms close to urban areas, where tearooms and farm shops get lots of customers.
- The set-up costs can be high and there is no guarantee of success.
- Increasing numbers of visitors and traffic irritate local people.

Many farmers have turned to organic farming as a way of diversifying. This means growing crops and rearing animals without the use of chemicals (pesticides, herbicides, antibiotics). Organic food is more expensive so profits can be higher. There is demand for natural and healthy food by the public:

- About 4% of farmland in the UK is organic.
- Organic food is often sold at farmer markets, cutting out the middle man so farmers get more profit.
- Organic food sales fell by 13% in 2009 because of the recession.
- Organic yields are often 50% lower than non-organic so prices have to be high to make a profit.

Check your understanding

Give two examples of farm diversification.

Tested

Exam practice

13 Describe the different ways farmers in the developed world can diversify. [4]

14 Explain what is meant by organic farming. [2]

Tested

Answers online

Knowing the basics

Farming is in decline – this is the basic cause of farms seeking to diversify.

Stretch and challenge

Farming jobs are often poorly paid and are sometimes both part-time and/or temporary.

Fair trade and intermediate technology

Revised

Helping developing world farmers needs a different approach to that taken in the developed world. Often the problems are very basic ones like soil erosion, lack of water, poverty and lack of food supply. For farmers of cash crops, **fair trade** is one way they can be helped:

- Coffee, cocoa and tea farmers join a fair trade co-operative to pool their farm produce to sell.
- Fair trade pays farmers a small extra amount of money for what they produce.
- Farmers also get a guaranteed minimum price, protecting them if world market prices fall.
- The extra money goes to farmers while some is spent on village facilities.

Kuapa Kokoo in Ghana is a Fairtrade cocoa farmers' co-operative set up in 1993 by a group of cocoa farmers who saw the opportunity to work together and ensure better livelihoods for the members. There are now 65,000 members from 1400 villages. Five per cent of Ghana's cocoa is produced by *Kuapa Kokoo*. The co-operative gets a guaranteed minimum price of US$2000 per tonne of cocoa plus an extra US$200 Fairtrade premium. The premium is invested by the farmers and from it each farmer gets a bonus payment each year, and village projects are financed including wells for clean water, mobile health clinics and schools.

Criticisms of fair trade include the fact that consumers in developed countries believe they have to pay more for Fairtrade products and they are not always prepared to do this. However, through Fairtrade, farmers are not only receiving extra income individually but investing their own money in community projects that each benefit hundreds of people. Fairtrade also means that farmers are being trained in better farming methods, and women particularly are learning additional income-generating skills as cocoa farming is seasonal.

For subsistence farmers, one way forward is the use of **intermediate technology**. This is simple technology that is easy and cheap to build and maintain. Local people can be taught the skills needed to construct and use it. The NGO Practical Action funds intermediate technology farming projects for subsistence farmers in northern Sudan:

- **Zeer pots**
  These are a clay 'pot inside a pot' with the space between the pots filled with damp sand. Women are trained to make the zeers which can keep vegetables fresh for up to 20 days. This reduces food waste, improves food supply and health. Women can also earn money by making and selling their zeers

- **Crescent terraces**
  These are small banks of earth and stones heaped up along the contours of slopes. When it rains, the banks stop water runoff and trap any eroded soil. This reduces soil erosion and encourages water to infiltrate – producing wetter, richer soil which then increases crop yields

- **Earth dams**
  Practical Action trains villagers to build small earth dams across streams that only flow in the wet season. These store water for the dry season and the water can be used to **irrigate** crops

- **Animal-drawn ploughs**
  Traditionally, farming was done with a hand-held hoe. Local blacksmiths have been trained to make simple wood and metal ploughs that can be pulled by cattle or even camels. This means larger area can be ploughed, increasing yields

Intermediate technology has improved food and water supply in northern Sudan, reduced erosion and given subsistence farmers ways to grow more and earn more. However, these projects would need to be expanded enormously to assist all the subsistence farmers in the world who need help.

Knowing the basics

Fair trade provides farmers with extra income which could lift them out of poverty.

Stretch and challenge

Fair trade products have become popular, but because they are more expensive, sales declined in the recent recession because people switched back to cheaper alternatives.

Exam practice

Tested

15 Describe the advantages and disadvantages of fair trade. [4]

16 Using named examples, explain how intermediate technology can be used to help farmers in the developing world. [8]

Answers online

Check your understanding

State two features of fair trade.

Tested

Making Geographical Decisions

Key Idea 1: Sustainable development is an important concept

The question of deciding whether something is 'sustainable' or not crops up a lot in geography. Often the question occurs when a new project is being developed, such as The Three Gorges Dam from Unit 2 (Chapter 12).

Sustainable development was defined in the United Nations World Commission on Environment and Development (WCED) report 'Our Common Future' (1987). The definition below is often called the 'Brundtland' definition, after the chairperson of the WCED, Gro Harlem Brundtland:

'Sustainable development is development that meets the needs of the present without compromising the ability of future generations to meet their own needs.'

This definition suggests that **human development** is sustainable when:

- it meets the needs of people today, i.e. incomes and quality of life improve
- development today does not reduce quality of life for people in the future.

One way of deciding whether development is sustainable is to use the stool of sustainability model (Figure 1). This uses three criteria (social, economic and environmental) to judge whether development is sustainable. If one of the legs is judged to be unsustainable then the stool will 'fall over'.

Figure 1 The stool of sustainability

Key features	Sustainable development	Unsustainable development
Social	• Directly improves life, such as better education, health and housing • Promotes equality between different groups	• Excludes some people, such as women or an ethnic minority group, increasing inequality • Imposed on people from above, so they have no say in decision-making
Economic	• Increases income for many people long term, so quality of life improves • Proves to be good value for money, benefiting the most people for the least cost	• Leads to debt, which future generations will have to pay back • Only makes a small number of people wealthy and increases inequality
Environmental	• Helps conserve **biodiversity**, e.g. selective logging rather than large-scale **deforestation** • Uses non-polluting and **renewable resources**	• Wastes finite **natural resources**, e.g. fossil fuels, which are then not available in the future • Causes **pollution**, e.g. industry polluting the air and water, that future generations will have to clean up

Check your understanding

Tested ☐

What is the definition of sustainable development?

Knowing the basics

Sustainable development is important because what humans do today should not make life harder for future generations.

Questions to think about:

1 Re-examine the Three Gorges Dam project on page 94. Do you think this project was economically, socially and environmentally sustainable?

2 Are bottom-up development projects (page 93) more sustainable than top-down ones?

3 Is there such as thing as a sustainable level of population? (See Unit 2, Chapter 9.)

Stretch and challenge

In reality, no development is 100 per cent sustainable. There will always be some negative impacts on the environment and/or some people. The trick is to try and minimise these negative impacts as far as possible.

Key Idea 2: Since the 1990s 'environmental sustainability' has become increasingly important

Since the 'Our Common Future' report in 1987, the word 'sustainability' has increasingly become to mean 'green' or 'eco-friendly'.

This is possibly because since the 1980s environmental issues have become much more important in terms of public awareness, such as:

● **climate change** and **global warming** (Unit 1, Chapter 2, Changing Climate)

● biodiversity loss and deforestation (Unit 1, Chapter 3, Battle for the Biosphere).

At the same time, many people have experienced levels of human development because of **globalisation** (Unit 2, Chapter 11) and development projects (Unit 2, Chapter 12).

This is a definition of environmental sustainability from the business website SmallBizConnect:

'Environmental sustainability involves making decisions and taking actions that are in the interests of protecting the natural world.'

http://toolkit.smallbiz.nsw.gov.au/part/17/86/371

Notice that this definition is about the natural environment rather than human development (the definition of sustainable development on page 131 refers to human development).

Different organisations have different attitudes towards environmental sustainability:

TNCs such as BP and VW TNCs often stress their environmental policies and efficiency, but are sometimes accused of '**greenwashing**' and '**tokenism**' by environmentalists – using slick marketing and public relations to appear more environmentally friendly than they really are.	**NGOs such as Oxfam and CAFOD** Many **NGOs** that work in the developing world emphasise the need to use small-scale, **intermediate technology** such as water hand-pumps, solar panels and **rainwater harvesting**. They focus on using renewable resources.
Conservation organisations such as the WWF The World Wide Fund for Nature (WWF) focuses on the eco-footprint concept. This suggests humans are already 'living beyond their means' in terms of using natural resources.	**Environmental pressure groups such as Greenpeace** Pressure groups such as Greenpeace focus on pollution and waste, such as the damage done to the environment by industry, energy **consumption** and the 'throw-away' society.

It is important to realise that people and organisations have very different attitudes towards the need for environmental sustainability:

● Some see it as a marketing opportunity. By appearing 'green' and 'eco-friendly' consumers will buy more products and profits will be higher.

● Others believe that humans can lead very similar lives to the today, but only if we switch over to renewable energy resources, recycle more and use resources more efficiently.

● A radical alternative is what is called a 'no-growth' or zero-growth economy; humans should stop trying to get ever more developed and restrict economic and population growth.

These views about environmental sustainability are very different. It would be very difficult to get all organisations and people to agree.

Check your understanding

Tested ☐

How is 'environmental sustainability' different from 'sustainable development'?

Questions to think about:

1 Look back at the energy resources you studied in Unit 2, Chapter 10 and consider if they are environmentally sustainable.

2 Think about trade and globalisation from Unit 2, Chapter 11. Is global trade good or bad for the environment?

3 Identify the management measures from Unit 1, Chapter 3 that are designed to make humans' relationship with the biosphere more environmentally sustainable.

Knowing the basics

In terms of sustainability, the shift from a focus on human development towards environmental issues happened in the 1990s.

Stretch and challenge

Many people are very critical of the concept of environmental sustainability, because it reduces the complex idea of 'sustainable human development' down to a narrow focus on not harming the environment.

Key Idea 3: Demand for resources is rising globally but resource supply is often finite, which may lead to conflict

The main reason that sustainability has become such a buzzword is because, increasingly, the supply of resources does not match demand:

Resource demand	Resource supply
<ul><li>World population was 3 billion in 1960, grew to 7 billion in 2010, and could be 10 billion by 2050</li><li>Energy and water demand is rising fast and could double by 2035</li><li>Demand for food could also double between 2010 and 2050</li></ul>	<ul><li>Over 80% of global energy demand is met by fossil fuels, a finite resource</li><li>Some resources are showing signs of reaching peak supply, such as oil, wild ocean fish</li><li>The amount of 'spare' land that could be used for farming is very small</li></ul>

The constant demand for more resources leads to:

- overuse of water supplies, so they dry up and become polluted by farming and industry
- deforestation, which means some species are lost
- increased levels of CO_2 and other gases in the air as more fossil fuels are burnt.

A key issue is growing affluence (increasing wealth). As countries develop and people get richer, they:

- eat more meat, fat and sugar and less carbohydrate and fibre – meat and processed foods take more land and energy to produce
- move into larger, better serviced houses in urban areas – and begin to use electricity, piped water and other services
- become more mobile, and transition occurs from bicycles to motorbikes and cars – which burn fossil fuels
- demand products, like foods, from around the world – they become part of globalisation and global trade, and use ever more resources.

This means that the wealth and development level of people is just as important as the number of people in terms of resource consumption.

Conflict over Nigeria's oil

Oil was discovered in Nigeria, in the delta of the Niger River, in 1956. Oil is extracted by the TNCs Shell, Exxon-Mobil, Chevron and Texaco, but has caused continuous problems in the Niger Delta:

- Shell has been accused of allowing thousands of barrels of oil to leak and pollute the sensitive delta ecosystem and farmland.
- Nigeria earns US$10 billion from oil each year, but almost none of this goes to the people in the delta who live in poverty.
- The Ogoni people, who live in the delta, say they have suffered violence, evictions and health problems because of the oil extraction. NGOs like Amnesty International and Friends of the Earth have tried to raise awareness of their plight to the world.

Governments and TNCs want to profit from natural resources, but this often leads to environmental problems and inequalities which local people and NGOs see as unfair and destructive.

Check your understanding

Tested

An increasing population causes rising demand for resources. Why does increasing wealth do the same?

Knowing the basics

Many natural resources have a finite supply; once we use them up there will be no more.

Questions to think about:

1 How useful would population policies, like China's one-child policy (Unit 2, Chapter 9), be for reducing resource demand on a global scale?

2 Would there be more or less conflict if humans developed renewable resources (Unit 2, Chapter 10) rather than non-renewable ones?

3 What is the link between rising resource demand and climate change (Unit 1, Chapter 2)?.

Stretch and challenge

As the world becomes richer and more globalised, even very isolated extreme environments are likely to come under increased pressure in terms of resource exploitation.

Key Idea 4: Balancing the needs of economic development and conservation is a difficult challenge

Governments have a very difficult balancing act when it comes to **economic development**. They need to ensure the economy of a country can provide enough income so people can have a good quality of life, but they need to prevent serious pollution and land degradation which destroys the natural environment. This is the tension between economic development and conservation (Figure 2).

Figure 2 Economic development or conservation?

Conservation is the priority   Economic development is the priority

If economic development is prioritised:

- land will be lost to **urban sprawl**, factory and transport developments
- pollution regulations will be weak, leading to environmental degradation and poor human health
- farming will be intensive and cause soil **erosion**, land and water pollution.

If conservation is prioritised:

- protecting **greenbelts** could lead to housing shortages
- money spent on conservation could be spent on social services
- overly strict regulations could prevent investment in factories and offices.

As environmental concerns have become global rather than just national issues, global organisations like the United Nations have stepped in to put forward global solutions. Getting many countries to agree is very difficult, as the 1997 Kyoto Protocol proved:

The Kyoto Protocol

This was the first attempt at a global agreement to reduce **greenhouse gas** emissions. In Kyoto, Japan, in 1997 developed countries attempted to reach agreements. EU countries signed up to reduction targets, e.g. the UK agreed to a 12 per cent cut and Denmark to a 20 per cent cut, but:

- Canada agreed to a 6 per cent cut, yet withdrew from the agreement in 2011 and said that trying to cut fossil fuel use was damaging its economy.
- The USA and Australia did not sign up to any cuts in 1997 because they believed this would close down factories and power stations, and jobs would be lost.
- some countries such as Spain have ignored the target they agreed to, whereas Finland has reduced emissions beyond its target.

The Kyoto Protocol shows that countries have very different attitudes and priorities. It is important to realise that when elections take place and governments change, policies change too. Australia's government today is more willing to reduce emissions than the government in power in 1997. In some Scandinavian countries, such as Finland and Denmark, green political parties are more powerful than in other countries.

Check your understanding
Tested

What is the Kyoto Protocol?

Knowing the basics

Balancing the need to conserve the environment with the need to develop the economy is very difficult.

Questions to think about:

1 Can conservation, such as national parks and tropical forest reserves (Unit 1, Chapter 3), deliver economic development as well as conservation?

2 If humans continue to emit large amounts of greenhouse gases and this causes global warming (Unit 1, Chapter 2), could this have economic costs?

Stretch and challenge

Having clean air, water and land is actually very important for human health, as a polluted environment can cause many illnesses and health problems.

Key Idea 5: Achieving sustainable development requires funding, management and leadership

Even when a government has decided to adopt policies that favour sustainable development and environmental sustainability, there can be problems. New policies need to be funded (mostly from taxes) and managed so people are not negatively affected:

Wind turbines in the UK	Recycling in the UK
The UK has a target of getting 15% of its energy from renewable sources by 2015. Much of this will come from wind turbines. But installing wind turbines has met opposition: • Many people do not want to live near a wind turbine, saying they are ugly and noisy (it has become a 'NIMBY' issue – 'Not In My Back Yard'). • Coal and gas are cheaper sources of energy so the government has to provide subsidies to the wind power industry • As wind does not constantly blow, standby power stations will be needed	The UK has a target to reduce the amount of household waste going to **landfill** sites by 2020 to 35% of the amount in 1995. To meet this target, local recycling has to increase, but: • education is still needed to persuade the public that recycling is worthwhile • councils have had to spend money on new recycling bins for every household and new trucks • fines for not recycling were introduced but had to be scrapped, as the public was angry about them • many councils only collect bins once a fortnight to save money, which has angered some taxpayers.

For some important sustainability issues it is NGOs, pressure groups and environmentalists rather than governments that lead the way. They launch campaigns to try and persuade the public, governments and business to change their attitudes and actions:

● **Fair trade** has been promoted by the NGO Fairtrade International so successfully that even brands such as KitKat and Dairy Milk are now fair trade products.

● In the 1970s, Greenpeace and other NGOs began the 'Save the Whale' campaign to ban **commercial** whaling. Often direct action like ramming whaling ships was taken and made headline news.

● Chico Mendes was a Brazilian rubber plantation worker who became a globally famous environmentalist in the 1980s. He was one of the first people to highlight deforestation and the plight of **indigenous** people in the Amazon.

Check your understanding

Tested ☐

State two reasons why developing wind power in the UK has proved controversial.

Knowing the basics

Sustainable developments like wind turbines and recycling can be expensive to set up.

Questions to think about:

1 What actions would the UK government need to take if global warming (Unit 1, Chapter 2) meant sea levels rose by 1 m and rainfall decreased by 20 per cent?

2 How important is education (Unit 2, Chapter 10) in persuading the public that they need to lead more sustainable lifestyles?

Stretch and challenge

Even celebrity chefs like Jamie Oliver and Gordon Ramsay have played a role in bringing the issue of overfishing and the waste from fishing '**bycatch**' to the public's attention.

Key Idea 6: Physical processes and environmental changes increasingly put people at risk

Many people across the world are vulnerable to physical processes:

● Tectonic plate boundaries generate earthquakes, volcanic eruptions and sometimes tsunamis.
● Climate change is likely to mean rising sea levels, and possibly more floods and tropical storms.

In some locations, both tectonic hazards and climate change risks can be found in the same place. The coastal megacities of Asia are a good example:

Tokyo, Japan Population 36 million, growing at 1% per year	Manila, Philippines Population 12 million, growing at 2% per year	Jakarta, Indonesia Population 28 million, growing at 3.5% per year
Tokyo can be hit by tropical cyclones. The last major one was in 2007. Its port is built on reclaimed land at risk from sea level rise. It sits at the boundary of three tectonic plates, which is why major earthquakes occur. Its economy, US$1.4 trillion, could be devastated by a major earthquake	The city regularly experiences tropical cyclones as well as flooding and landslides. All of these could be made worse by global warming. Earthquakes are common because the city sits close to a plate boundary. About three million people live in slums, on very low incomes	40% of Jakarta is below sea level, and protected by **embankments** but very vulnerable to sea level rise. Thirteen rivers run through it so flooding is common – in 2007 400,000 people had to move due to floods. Under 50% of people in Jakarta have access to piped water. Many live in vulnerable slums and on less than US$2 a day

Many coastal areas are popular places to live. This is because many of the world's largest cities are on the coast and/or at the mouth of a flood prone river. These locations, chosen for ease of trade, are frequently close to plate boundaries and at risk from sea level rise and storms.

Managing the climate change and tectonic hazard risks in coastal cities is very difficult:

Demographic and social factors

● Population and building densities are high, and rising, so even accessing an area hit by disaster in the city, to help the injured, is almost impossible.
● Slum buildings easily collapse due to storm force winds, floods and earthquakes, and diseases like cholera can easily take hold in the aftermath of a disaster.

Economic factors

● In developing countries, city governments lack the funds to have large, modern emergency services.
● Many cities cannot afford complex sea or **flood defences** that might reduce the risk of climate change hazards.

The nature of hazard risk

● Some natural hazards, like earthquakes, cannot be predicted – they strike without warning.
● Predictions of global warming, and the increases in flooding and tropical cyclones, are intelligent guesses at best – it is hard to know how to prepare for an uncertain future.

The good news is that death tolls from major natural disasters have fallen, but the bad news is that the number of people affected, and economic losses, have risen.

Check your understanding

Tested ☐

How could future climate change affect rapidly urbanising cities in the future?

Questions to think about:

1 Which groups of people are most vulnerable to climate change and tectonic hazard risks (Unit 2, Chapter 12)?
2 Should governments plan for the maximum or minimum global warming projections you studied in Unit 1, Chapter 2?

Knowing the basics

Developing world megacities, with large concentrations of people on low incomes, are particularly at risk from tectonic and climate change hazards.

Stretch and challenge

It has been estimated that if a magnitude 9.1 earthquake, similar to the 2011 Sendai earthquake, struck Tokyo now the death toll would be 320,000.